Plastic Piping Systems

David A. Chasis
Plastic Piping Systems
Second Edition

Industrial Press Inc.

LIBRARY OF CONGRESS
Library of Congress Cataloging-in-Publication Data

Chasis, David A., 1938–
   Plastic piping systems / David A. Chasis. — 2nd ed.
     p.   cm.
   Bibliography: p.
   Includes index.
   ISBN 0-8311-1181-X
   1. Pipe, Plastic—Handbooks, manuals, etc. I. Title.
TH6330.C48 1988   88-18824
621.8'672—dc19   CIP

INDUSTRIAL PRESS INC.
200 Madison Avenue
New York, N.Y. 10016-4078

First Printing

PLASTIC PIPING SYSTEMS, SECOND EDITION

2 4 6 8 9 7 5 3

Contents

Preface

Plastic Piping Systems was first published in 1976. Amazingly, during the last decade very few significant changes have taken place in the plastic piping market. The commodity piping markets such as gas and water transmission, sewer and drain, drain waste and vent, acid waste and drain, sewer and water mains, and electrical conduit are still growing but, owing to the successful penetration of plastic piping in these markets, the rate of growth of plastic pipe use is much less today than it was 10 years ago. The only piping market that has not grown significantly with plastics is the industrial market. Estimates are that plastic pipe, fittings, and valves are capable of handling 65–70% of all the industrial applications in this market, but, yet, there are less than 5% of dollar volume purchases of plastic pipe, valves, and fittings. Why the lack of usage in this market?

There are several reasons. First, the marketplace is, as a whole, uneducated as to the "hows and whys" plastic piping products can handle many applications. The plastic piping industry has been remiss in educating the public of its product capabilities. What is needed is for the industrial plastic piping products manufacturers to unite and educate users, engineers, contractors, and students to the merits of designing with plastic piping systems.

Second, there has been a minimum of research and development in industrial plastic piping products. Most new products have been introduced by importers, with few new products or piping materials being developed domestically. There is a definite need for American innovation and research into plastic industrial piping products.

Third, industry is unwilling to change its habits and adapt to more progressive piping materials. This is especially true in such conservative manufacturing industries as pulp and paper, utilities, oil and gas, petroleum and ore refineries, and steel manufacturing. Europe and Japan dwarf the United States in the usage of industrial plastic piping products, and by using plastics they have effected millions of dollars of cost savings in all facets of their industrial process piping systems. American industry has to wake up and use the most effective and cost-saving piping materials it can to ensure its competitiveness worldwide.

Recently, two markets in the United States have become multimillion pound users of plastic piping: (1) The fire sprinkler market has been introduced to plastics (CPVC and polybutylene) to be used for home and commercial fire sprinkler systems, and (2) double-containment piping systems, handling toxic materials above- and below ground, have used several plastic piping materials. Federal, state, and local codes regulating these industries will encourage the use of plastic piping materials for both of these applications. Sanitary piping, used in a smaller and more specialized market, is starting to consider plastics. In the last few years a U.S. manufacturer has introduced PVC, polypropylene, and PVDF quick-disconnect-type piping systems for sanitary installations. This is a step in the right direction, but with added research and perhaps new material introduction incorporating higher temperatures and tensile strengths, stainless steel may be replaced as the primary piping material for sanitary applications.

The channels of distribution for plastic products have changed little in the last 10 years. The commodity products are handled by the same methods of distribution (manufacturer representatives and wholesalers) as the materials that they replaced. For example, plumbing wholesalers who had cast iron and copper DWV in the past are now stocking and selling ABS and PVC materials. Electrical distributors are converting portions of their inventory from metal to PVC conduit, and utility wholesalers that used to market cast iron, asbestos–cement, and other nonplastic materials are involved with PVC, RTRP, and polyethylene materials. In the industrial market plastic specialists still are the most significant factor in distributing thermoplastics and thermoset piping systems. My estimate is that 25–30 U.S. distributors handle up to 40–50% of the *industrial plastic* piping market. These distributors are unique in the marketplace in that in most cases they have undertaken the normal manufacturer's task of technologically educating and enlarging the marketplace. And they have! But as they have grown and prospered, their early pioneering efforts have waned, since they now spend more time managing their considerable businesses and less time educating and enlarging the market. When U.S. industry realizes the potential cost savings of plastic pipe, and the market increases in size two- or threefold, the plastic specialists will change or they will be replaced by the more conventional industrial supply houses.

The revisions in this second edition have been substantial. The Appendix has been expanded to include a comprehensive glossary, a listing of commonly used plastic and industry abbreviations, an expanded chemical resistance chart of over 600 chemicals, and many other useful items. Chapter I, Material Selection, is the most changed of all chapters in that there is included historical information of the plastics industry, a listing of the advantages of plastic piping systems, the provision of basic thermoplastic and thermoset manufacturing information, and additional information on flammability and toxicity of plastics. Few changes have occurred in Methods of Joining Plastics, Chapter III, but Chapter II, Piping De-

sign, has been increased to include much more information on designing both above- and below ground piping systems. Installation, Repair, and Maintenance, Chapter V, has not changed much, but Chapter IV, Product Selection, has been revised completely to include the updated piping material sizes as well as dozens of photographs of the various pipe, valves, fittings, and fabrications available today. Chapter VI, Cost Comparisons, has been updated and Chapter VII, Applications, has been expanded greatly, both pictorially and verbally. This Second Edition of *Plastic Piping Systems* is more new than revised. I wanted to include the information you needed and could use easily. I hope I have accomplished this. Please write me with any comments and suggestions you have to make the next revision better.

I would like to express my appreciation and gratitude to the many plastic associations, distributors, and manufacturers that contributed to this revised text. I have attempted to credit where applicable all contributions of technical data and pictures. All technical data shown in this book have been used directly or extrapolated from existing manufacturers's published literature. The information presented is the best available and can be used as an excellent guide. However, if there is any extremely critical or marginal applications, I strongly recommend that you check directly with the manufacturer before deciding on the use of a particular plastic material. Although the material published has been checked and rechecked several times, errors may exist. If so, I bear that responsibility.

Besides the support given to me by the plastics industry, I would like to acknowledge the very professional and capable assistance of two word processor operators, B. J. Derton and Kathi Estes, CPS. Special thanks go to my very close friend Dennis Garber, without whose prodding and suggestions the book would have had less content and usefulness. Finally, without the love and inspiration of my lovely daughters, Hara, Rebecca, and Gabrielle, this revision would not have been as enjoyable an experience or as likely to have been finished.

Plastic Piping Systems

1

Material Characteristics

HISTORY

The plastics industry can trace its auspicious beginning to the scarcity of ivory in the manufacturing of billiard balls. John Wesley Hyatt, a creative and inventive chemist, mixed pyroxylin (a derivative of cotton) and nitric acid with camphor to form a product he called celluloid.

Like so many other technological discoveries, war became the catalyst necessary to promote the research and development of plastics. During and just after World War II, polyvinyl chloride, polyethylene, and reinforced plastics were introduced into many industries; but, they were not introduced into the piping industry until the late 1940s. Germany and Japan embraced plastics, especially thermoplastics, since it allowed those war torn countries, with a minimum investment of capital and other resources, to manufacture much of their piping needs quickly and economically. Germany and Japan are still leaders in volume producing and technology within the thermoplastic piping industry. (Table 1-1 lists the dates of discovery of commonly used plastics.)

In the United States, the usage of plastics in piping started slowly. It did not obtain wide public acceptance until the late 1950s and early 1960s. Since then, thermoplastic and thermosetting pipe usage has increased at an astounding rate. (The reasons for the meteoric growth of plastic piping are explained later in this chapter under "Advantages".)

Whereas metal and other nonplastic piping products have grown at a rate of less than 2% in the past 10 years or so, plastic piping has approached four times the growth rate of nonplastic

piping. The estimated market for all plastic piping, fittings, tubing, and conduit is about 4 billion pounds of product worth $2.7 billion. Table 1-2 lists pipe resin usage by material for 1986.

The underground piping market is the largest use segment of plastic pipe. Potable water distribution, drain/waste/vent, sewer and drain, gas, irrigation, conduit, and pressure pipe are by far the largest markets for plastic piping not only in the United States but in the world. Process piping accounts for less than 5% of the plastic piping market in dollars and approximately 15–20% in footage.

ADVANTAGES OF PLASTIC PIPING

The marketplace's spectacular acceptance of plastics for piping applications is due to the many desirable features of plastics compared with other materials. These features offer many advantages that effect considerable cost savings and an increase in piping system reliability. The following paragraphs list the many advantages of plastic piping systems.

Corrosion Resistance—Plastics are nonconductive and are therefore immune to galvanic or electrolytic erosion (a major cause of buried metal pipe failure). Because the outer wall of the plastic pipe is so corrosion resistant, plastic pipe can be buried in acid or alkaline and wet or dry soils with no paint or special protective coating applied.

Chemical Resistance—The variety of materials available allows almost any chemical, at moder-

Table 1-1. Discovery Dates of Commonly Used Plastics

Date	Material	Date	Material
1868	Cellulose nitrate	1945	Cellulose propionate
1909	Phenol-formaldehyde	1947	Epoxy[a]
1919	Casein	1948	Acrylonitrile-butadiene-
1926	Alkyd		styrene[a]
1926	Analine-formaldehyde	1949	Allyic
1927	Cellulose acetate	1954	Polyurethane or
1927	Polyvinyl chloride[a]		urethane
1929	Urea-formaldehyde	1956	Acetal
1935	Ethyl cellulose	1957	Polypropylene[a]
1936	Acrylic	1957	Polycarbonate
1936	Polyvinyl acetate	1959	Chlorinated polyether[a]
1938	Cellulose acetate	1962	Phenoxy
	butyrate[a]	1962	Polyallomer
1938	Polystyrene or styrene[a]	1964	Ionomer
1938	Nylon (polyamide)	1964	Polyphenylene oxide
1938	Polyvinyl acetal	1964	Polymide
1939	Polyvinylidene chloride[a]	1964	Ethylene-vinyl-acetate
1939	Melamine-formaldehyde	1965	Parylene
1942	Polyester[a]	1965	Polysulfone
1942	Polyethylene[a]	1970	Thermoplastic polyester
1943	Fluorocarbon[a]	1973	Polybutylene[a]
1943	Silicone	1975	Nitrile barrier resins

[a]Plastics used for piping materials.
Source: Adapted from SPI literature.

Fig. 1-1. Typical corroded and pitted metal piping after being buried for a short period. (Courtesy of R&G Sloane Manufacturing.)

ate temperatures, to be handled successfully by plastic piping. Table 1-5 lists the chemical resistance of plastic piping materials in chemical groups, and in appendix B there is a more extensive listing of over 600 chemicals.

Low Thermal Conductivity—All plastic piping has low thermal conductance. This feature maintains more uniform temperatures in transporting fluids in plastic piping than metal piping. Minimal heat loss through the pipe wall of plastic piping may eliminate or reduce greatly the need for pipe insulation.

Flexibility—Thermoplastic piping materials are relatively flexible. Piping flexibility is a major asset particularly in underground piping installations. Plastic piping may be "snaked" in trenches to minimize the effects of expansion and contraction; underground pipe bends may be used more readily with plastics eliminating or minimizing the use of pipe-fitting ells and ensuring a more dependable piping system; the very effi-

Table 1-2. Pipe Resin Usage in 1986 (10^6 lb)

Polyvinylidene chloride (PVC)	3110
Polyethylene	680
Acrylonitrile–butadiene–styrene (ABS)	160
Reinforced thermosetting resin pipes (RTRP)	140
Chlorinated polyvinyl chloride (CPVC)	10
Polypropylene (PB)	5
Polybutylene (PB)	3
Other	2
Total	4110

Source: Adapted from *Plastics World* (May 1987 issue).

cient "plowing" installation technique may be used, owing in part to plastic piping's flexibility and toughness; insert pipe renewal of deteriorated metal or concrete pipe with plastics is possible owing to the flexible nature of plastic piping.

Low Friction Loss—The interior walls of all plastic piping have a Hazen and Williams C factor of 150 or higher. In most cases, this means less horsepower is required to transmit fluids in plastic piping compared with metal and other nonmetallic piping. This feature plus the excellent corrosion resistance of plastics allows smaller diameter piping to be used in place of larger nonplastic piping systems with a cost savings to the user.

Long Life—Owing to the minimum effects of internal and external corrosion on plastics, the excellent weatherability of plastics, and their relative inertness, there is very little change in the physical or molecular characteristics of plastic piping over dozens of years. Upon examination of several plastic applications that had been installed a minimum of 25 years ago, no measurable effect of any degradation was found. Theoreti-

A

B

Fig. 1-2. (a) The smooth interior of plastic piping allows less friction loss of transported fluids. (Courtesy of PPS-Maryland.) (b) Plastic piping showing glossy-smooth inner bores.

cally, in most installations of plastic piping, there is no known end-life of the piping system.

Lightweight—Most plastic piping systems are a minimum of one-sixth the weight of steel piping. This feature means less freight cost owing to being able to load more lengths of plastic piping on a vehicle than metal. This feature also greatly reduces installation cost. It is much easier to install lighter-weight piping systems when piping in close quarters; and, in many cases, underground piping requires no expensive heavy lifting equipment. Polyolefin piping is the lightest of plastic piping materials and floats on water. This characteristic has allowed some very unique and cost saving installation procedures in several marine applications.

Variety of Joining Methods—Plastic piping can be cemented, heat-fused, threaded, flanged, and compression fitted (insert, rolled grooved, compression nut, flared, and O-ring coupled). This variety of joining methods allows plastic piping to be adapted easily to any field application.

Nontoxic—Plastic piping systems are nontoxic and odorless. Several plastic systems are listed as approved materials with the National Sanitation Foundation.

Biological Resistance—To date there are no documented reports on any fungi, bacteria, termites, or rodent attacks on any unplasticized thermoplastic or fiberglass-reinforced piping. In fact, owing to its inertness, smooth interior, and resistance to biological attack, plastic piping is the preferred material in deionized and other high-purity water applications.

Abrasion Resistance—Thermoplastics and thermoset piping materials have provided excellent service in handling slurries such as fly ash, bottom ash, and other abrasive solutions. The molecular toughness and inner-bore smoothness of plastic piping make it ideal for abrasive-resistant applications.

Complete System Material Integrity—Complete systems exist in PVC, CPVC, polypropylene, PVDF, Teflon, and RTRP materials including pipe, fittings, valves, tanks, pumps, air-handling equipment, and tubing. This feature allows a plastic piping material to be incorporated entirely as the carrying medium for a particular hard-to-handle fluid.

Weather Resistance—With the use of additives in base resins by the piping manufacturer, plastic piping systems are superb materials in any nat-

Fig. 1-3. (a) An example of plastic pipe's light weight is this installer carrying 100 ft of 2 in. RTRP at a construction site.

ural environment. For example, PVDF has such good weather resistance that it is used as the finished coat on outdoor metal-clad roofing. Also, owing to the elongation and recovery properties of polyolefins and styrene piping systems, freezing and thawing will not significantly damage the integrity of the piping system.

Code Uniformity—Owing to the efforts of manufacturers and standards-issuing agencies, there are dozens of plastic piping standards on ther-

Fig. 1-3. (b) Twelve-inch RTRP ceramic embedded construction handling mine slurry. (Courtesy of Smith Fiberglass Products.)

moplastic, thermoset, and plastic-lined piping materials. These standards ensure products that have uniform characteristics and, in most cases, allow each manufacturer's materials to be used with another manufacturer's product. This feature ensures product availability and competitive pricing to the *end user.*

Domestic Manufacturered Products—The United States has borrowed heavily from European and Japanese thermoplastic technology with the result being that all commonly used plastic piping materials are quality manufactured in quantity domestically. The United States dominates the international thermosetting piping industry in technology and material availability. This feature ensures excellent product availability with immediate technical service backup and constant prices (not depending on fluctuations of foreign currency).

Colored Piping—The plastic piping manufacturing process allows any color pipe to be made. Usually, natural, blacks, whites, and grays are used in most piping systems. For underground piping, the gas industry uses bright orange and tan plastic piping and the water industry uses light blue, green, and white piping. These vibrant colors are easily seen when contractors are excavating, preventing pipe damage and saving lives as well as thousand of dollars in repairs. With other piping materials, bright ribbons or dyes or no markers at all are used.

Maintenance Free—A properly installed piping system requires *no* maintenance. There is no rust, pitting, or scaling. The interior and exterior piping are not subjected to galvanic or electrolytic corrosion. Painting, coating, and other preventative maintenance is *not* required. In buried applications, the plastic piping is *not* affected by even the most aggressive soil conditions.

Energy Conservation—Cumulative energy requirements to manufacture and install plastic piping materials are 50–70% less per unit volume than for steel and aluminum piping. This includes energy (Btu/in.^{3}) consumed in manufacturing, transportation, installation, operation, maintenance, and fuel value of the product itself. Added to this is the savings in energy (horsepower) required to move fluids in plastic piping systems versus other nonplastic piping systems.

Lower Overall Costs—When all the previously listed features of plastic piping are considered, it requires substantially less cost to use plastic piping systems; these features are the main reason why plastics are being used more frequently by the piping industry. In chapter 6 a more detailed cost comparison will be made of plastics versus nonplastic piping materials.

PLASTIC PIPING MANUFACTURING

This book is not intended to make everyone a plastic piping manufacturing expert. However, it is useful to know something about the processes involved in taking a pure polymer to the usable piping end product. With this knowledge, a respect for the chemical and mechanical expertise needed to produce a rather simple looking 20 ft length of 4 in. plastic pipe is gained. Also, by reviewing the pipe and fitting manufacturing processes, the major differences between thermoplastic and thermoset piping products will be easier to discern.

THERMOPLASTICS

Almost all thermoplastic piping is extruded and all thermoplastic fittings are injection molded (or fabricated from extruded pipe). This commonly shared manufacturing technology allows thermoplastic products such as PVC 80, CPVC 80, PVC 40, PVC DWV, polypropylenes, and polyethylenes to have uniform qualities and physical compatibility, so that one manufacturer's standard PVC 40 fitting may be used with any other manufacturer's standard PVC 40 pipe or fitting. This important feature, unfortunately, is not necessarily true in reinforced thermosetting resin pipe products.

Extrusion

Extrusion is defined as the forcing of a melt through a shaping die. In plastic piping extrusion, solid materials in the form of powder, flakes, pellets, beads, or granulated regrind mixed with additives (see Table 1-3) are fed into the extruder. The material is conveyed to the die by one or more screws enclosed in a long cylindrical barrel. The resinous material is then subjected to high temperature and pressure, which forces the resin melt through the die at a predetermined rate. (See Fig. 1-4.) After the pipe is formed, it is cooled and pulled to a winder or a cutoff device. The piping production rate is normally limited to the speed at which the pipe can be sized and cooled properly. In order to prevent internal strains and profile distortion, cooling and annealing of the piping profiles must be done carefully.

Injection Molding

Similar to the extrusion process, granular or powdered resins are fed into a machine where they are transported, melted, mixed, and formed in a mold (not a dielike extrusion). When the plastic cools and solidifies in the shape of the mold cavity, the mold opens and the parts are ejected; these parts require little or no postfinishing. Most injection molding machines are horizontally configured, but some vertical machines are in use. Both horizontal and vertical machines operate on similar principles, except the vertical machine uses gravity to hold the mold inserts. Molding machines consist of two basic components, a clamp unit with a stationary platen and a movable platen, which mounts the mold. This combination holds the mold and clamp together under pressure, allowing the molten plastics to be injected into the cavities. When the melt cools and solidifies, the clamp opens and the finished part is ejected. The other basic component of injection molding machines is the injection end. The plastic is fed into the plasticating unit where it is heated and melted and is forced into the closed mold cavities under high pressure. (See Figs. 1-5 and 1-6.)

Fig. 1-4. Thermoplastic extruder showing pipe head attachment. (Courtesy of Cincinnati Milacron.)

Fig. 1-5. Three hundred and seventy-five ton hydraulic injection molder. (Courtesy of Cincinnati Milacron.)

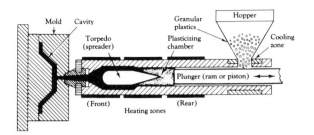

Fig. 1-6. Simple schematic diagram of plunger-type injection molding. (From *Composites: A Design Guide;* courtesy of Industrial Press Inc.)

Fig. 1-8. Application of a resin-rich inner layer showing surface C-veil in a filament-wound pipe. (Courtesy of Smith Fiberglass Products.)

THERMOSETS

Reinforced thermosetting resin piping and fittings are manufactured using four different processes: filament winding, centrifugal casting, contact molding (hand lay-up), and compression molding. There are distinct advantages to each manufacturing method. The inherent characteristics of each piping product formed by each process will be discussed subsequently in this chapter. The three most commonly used resins for all processes are epoxy, polyester, and vinylester.

Filament Winding

Filament winding is the most commonly used method of making RTRP, especially for large diameters. This process uses a continuous length of fiberglass filament (yarn or tape) wound onto the outside of a rotating mandrel. Mandrels may be composed of wood, steel, or aluminum and may be noncollaspible or collaspible. The collaspible type is more expensive, but greatly facilitates the pipe's release from the mandrel. The reinforcement may be saturated with resin or preimpregnated with partially cured resin. Specially designed machines wind the fiberglass at a manufacturer's specified pitch around the mandrel. Winding is continued until the desired wall thickness is achieved. Additional resin is then applied; this resin is composed of additives giving good external chemical resistance and ultraviolet protection. The resin is then cured by either endothermic (internal) or exothermic (external) reaction. The mechanical properties of the product may be altered by varying the pitch angle and the applied tension to the reinforcement. For use in corrosive environments, the winding may cover a resin-rich internal layer of glass mat or E-type glass. The final external layer for this severe service pipe will also have a resin-rich outer liner. (See Figs. 1.7 and 1.8.) For some pipe diameters exceeding 36 in. or when extra pipe stiffness is required (usually for buried large-diameter pipe), rib reinforcements may be affixed easily in this process to provide a stronger and tougher piping product. (See Fig. 1.9.)

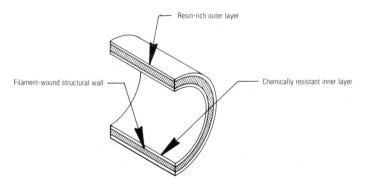

Resin-rich outer layer

Filament-wound structural wall

Chemically resistant inner layer

Fig. 1-7. Cross section of filament-wound pipe wall.

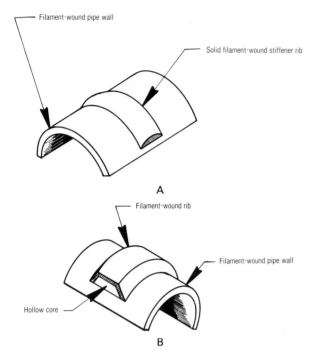

Fig. 1-9. (a) Half-elliptical rib-wall construction of RTRP reinforcement. (b) Trapezoidal rib-wall construction of RTRP reinforcement.

Centrifugal Casting

Centrifugal casting is a process that applies chopped glass fibers in a resin mixture to the inside of a rotating cylindrical mold. A cylindrically symmetrical pipe is formed as the resin cures. The strength properties of this piping material are isotropic (equal in all directions), owing to the orientation of the randomly applied glass fibers.

Centrifugal casting ensures a uniform and smooth outside diameter; filament winding provides piping product with a smooth uniform inside diameter.

Contact Molding

Contact molding is the method of fabrication that normally requires the fewest pieces of capital equipment and is incorporated as a preferred process by many small pipe fabricators for specially designed systems. Layers of reinforcing media (mats, rowing, etc.) are applied to a rotating mold or mandrel by hand. The desired resin is then sprayed or brushed on after each layer is positioned. The glass layers are applied, followed by resin impregnation, until the desired thickness is obtained. Heat may be applied to speed up curing, but usually room temperature is sufficient for curing. (See Fig. 1-10.)

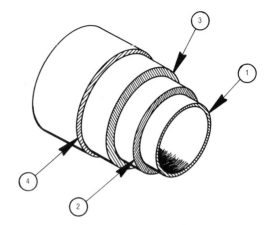

Fig. 1-10. Cutaway view of typical contact molded pipe: (1) smooth inner layer that is resin rich, with 90% resin and 10% glass; (2) next layer is reinforced with chopped strand mat and 25–30% glass; (3) additional layers are 70% resin and 30% glass; (4) exterior layer is similar to the inner layer, incorporating a resin-rich surface veil.

Compression Molding

Compression molding uses closely fitted male and female molds, and produces products whose properties are isotropic. Unmolded thermosetting material is injected, usually into the female cavity, and is cured using a pneumatic or hydraulic press. The molds are designed for minimum material loss, which results from some of the charge being pushed out of the cavity (overflow) during closing. Several RTRP manufacturers use this molding technique for manufacturing fittings.

PLASTIC PIPING RESIN ADDITIVES

All plastic pipe and fitting compounds include base polymers and additives. The additive materials enhance the base resin's properties so that the end product meets prescribed manufacturer's and national standards. A listing of the commonly used additives is given in Table 1-3. Although standards are maintained, there can be small mostly negligible variations of thermoplastic piping manufacturer's products owing to each manufacturer's unique formulation or processing techniques. However, significant variances occur among thermosetting pipe manufacturers owing to the many acceptable but diverse resin formulations and varying manufacturing methods. For these reasons, most RTRP systems of one manufacturer cannot be interchanged with other RTRP manufacturer's products.

Table 1-3. Additives in Plastic Piping Compounds

Type of Additive	Purpose	Benefit	Additive Material
Antioxidants	Inhibit or retard reactions promoted by oxygen or peroxides	Extends the polymer's useful temperature range and service life	Phenols Amines Phosphites Thioesters
Antistatic agents	Inhibit and remove static electrical charges of synthetic polymers	Allows plastic to be ungrounded and carry flammable materials easily ignited by a spark or electrical charge	Ammonium compounds Amines Esters Alcohols
Colorants	Pigments and dyes in encapsulated liquid or concentrated form are used to impart color to plastic resins	Color plastics any desired hue	Titanium dioxide Carbon black Inorganic oxides Chromates Cadmium Iron Molybdate
Coupling agents	Improves bonding of organic and inorganic materials	Upgrades mechanical and electrical properties of compounds; these properties are retained even after severe environmental aging; used in glass-fiber-reinforced piping products	Silane agents Titanates
Fibrous reinforcements	Added to a resin mixture to improve the properties of the resin	Fibers improve strength-to-weight ratio of resins and a host of other physical, chemical, and electrical properties. "E" and "C" glass are the most commonly used fibers for RTRP systems	Aramid Carbon Glass Hybrids
Nonfibrous property enhancers	Used for same purpose as fibrous additives but for different endproducts and manufacturing processes	Same benefits as fibrous additives; it improves mechanical, thermal, and electrical properties of base resins	Short fibers Flakes Spheres Irregular shapes of glass Carbon Oxides Epoxies Polyesters
Fillers and extenders	Added to resins to improved physical and electrical properties, as well as reducing the cost of higher-priced resins	Allow products to be more economically produced in the marketplace with no loss of quality	Talc Calcium carbonate Kaolinite Feldspar Silica Alumina trihydrate
Flame retardants	Added to resins to affect the burning rate of the resin in one or more of the following areas: (1) interface with combustion reaction; (2) make products of pyrolysis less flammable; (3) reduce combustion heat transfer; (4) reduce defusion rate of pyrolysis products	Used specifically for those plastics that support combustion, such as the polyolefins, ABS, and RTRP resins. In fact, one manufacturer advertises its product as "flame retardant polypropylene" (FRPP)	Aluminum Paraffins Antimony oxides Zinc Borates Miscellaneous Oxides Glycols
Foaming agents	Added to a polymer to form minute gas cells thoughout the product	Foamed plastics are lighter in weight, have improved electrical and thermal insulative properties, increased strength-to-weight ratios, and minimal shrinkage compared to nonfoamed plastics	Aliphatics Alcohols Ketones Aldehydes ABFA
Heat stabilizers	Added to prevent the degradation of resins due to heat and light	Allows plastics, in particular, PVC compounds, to be stable and retain physical properties in excessive heat and light	Calcium Zinc Cadmium Lead Phosphites Organotins
Lubricants	Improve the processing of resins by lubricating the resin internally and externally	Improves processing time, allowing for quality products at reduced costs	Esters Fatty Acids Glycerols Polyesters

Table 1-3. (*Concluded*) Additives in Plastic Piping Compounds

Type of Additive	Purpose	Benefit	Additive Material
Lubricants (*cont'd*)			Metal soaps Paraffins Polyethylenes
Organic peroxides	Are a source for initiating both polymerization and copolymerization of vinyl and diene monomers	Speeds up and ensures the completion of polymerization and copolymerization, thus reducing the cost of resins; used in PVC, polyester, polyolefins, and cross-linked polyolefins	Organic peroxides
Plasticizers	Aids plastics to exhibit greater resilence, flexibility and impact strength by reducing the glass transition temperatures below the anticipated use temperature	Allows products such as vinyl tubing, vinyl flooring, and wire and cable insulation to be produced with excellent wear properties; plastic piping is unplasticized.	Di-phthalate Glycols Triacetin Epoxies Polyesters Sebacates
Preservatives	To prevent degradation of polymers by microorganisms	The addition of preservatives prevents fungi and bacteria attack on plastic piping, making the piping material ideal for underground installations and for handling "pure" liquids; prevents decontamination of piped fluids	Pesticides Biocides
Processing aids	To increase the effectiveness and efficiency in manufacturing processes; five common process aids are: (1) viscosity depressants; (2) mold release agents; (3) emulsifiers; (4) slip agents; (5) antiblocking agents	Five common benefits of these aids are: (1) decrease viscosity of a liquid resin; (2) ease release of products from molds; (3) lower interfacial tension between two or more liquids; (4) provide surface lubrication during and after processing; (5) prevents two surfaces from permanently bonding to each other	Mica Fatty acids Plastisols Anionic aminos Nonionic ethers Waxes Polyolefins
Smoke suppresants	Reduces the amount of smoke during combustion	Reduces the quantity of smoke in plastics; hence; could save property and lives	Molybdenum Zinc Iron compounds
Ultraviolent stabilizers	Protects substrates by retarding the degradative effects of sunlight	Allows plastic piping, particularly the olefins, to be used outdoors with no significant changes of physical properties	Carbon black Benzophenones Benzotriazoles Nickel organics Amines Benzoates Piperidines

DEFINITIONS OF MATERIALS

"Plastic" pipe is as vague a term as "metal" pipe. Like metals, many plastic piping materials are produced that exhibit a wide range of physical properties and characteristics. The properties of plastics differ depending on the organic structure of the base polymer(s) and, to a large degree, on the quantity and type of resin additives. Using PVC as an example, by varying the amounts and type of additives it is possible to produce an opaque gray, extremely strong, and rigid unit of pipe or a clear, soft, pliable product such as a plastic seat cover.

There are two basic types of plastics: thermoplastic and thermoset. Thermoplastic resins can be repeatedly heated to a liquid state and then rehardened by cooling with no measurable change in the resin's properties. A good analogy of thermoplastics would be water. Upon cooling, water will solidify to form ice, and, when heated, will transform to a liquid state. This process can be continuous with no chemical change. Piping materials, such as polyvinyl chloride (PVC), chlorinated polyvinyl chloride (CPVC), polyethylene (PE), polypropylene (PP), polybutylene (PB), acrylonitrile–butadiene–styrene (ABS), and fluorocarbons (PVDF, PFA, TFE), fit the thermoplastic classification.

Thermosets, however, once they are cured, cannot again be changed in shape with reheating without a physical and or chemical change to the resin. A good analogy of thermosetting resins would be a hard-boiled egg. Once a hard-boiled egg is formed, reheating will not melt and reform the egg. All reinforced thermosetting resin pipes (RTRP) are thermosetting materials.

A short summary and description of distinctive characteristics of commonly used plastic piping materials follows.

THERMOPLASTICS

Polyvinyl Chloride (PVC)—PVC is a piping resin prepared by the polymerization of vinyl chloride with minimum quantities of additives; it does *not* contain plasticizers. [In the past (and continuing in Europe) U.S. pipe manufacturers listed their pipe resin as UPVC, meaning unplasticized PVC.] Three types of PVC piping materials are produced. In this text, all charts and tables for PVC piping will be characteristic of PVC type I, grade I (cell classification—12454-B). Type II PVC (cell classification—14333-D) has higher impact strength than type I, but has reduced physical properties in other areas. Type III PVC (cell classification—13233) contains fillers that increase the stiffness but lower the tensile and impact strengths and reduce the chemical resistance. Some non-pressure piping, such as conduit and sewer and drain pipes, are made from this compound. The maximum service temperature of PVC is 140° F (60° C). Owing to its excellent long-term strength, higher stiffness, resistance to a broad range of chemicals, and competitive price, PVC comprises over 85% of the footage of all thermoplastic piping. Because of this popularity, PVC has the broadest range of valves, fittings, and other piping appurtenances of any plastic piping material.

Chlorinated Polyvinyl Chloride (CPVC)—CPVC is a resin made by postchlorination of PVC. The additional chloride in the PVC polymer allows the maximum service temperature of CPVC to be extended from 70 to 210° F (from 21 to 99° C). The material designation of CPVC is type IV, grade I (cell classification—23477-B). For all practical purposes, other than service temperature, CPVC has almost the same physical properties as PVC. CPVC is approximately twice as expensive as PVC, but, when higher temperatures are involved, it has proven to be an ideal material.

Polyethylene (PE)—PE is a resin prepared by the polymerization of ethylene, and is the second most widely used thermoplastic piping material and the best known of the polyolefin group of plastics. The polyethylene polymers normally have two major additives: antioxidants and pigments.

The antioxidant assists in processing, and the pigmentation (normally carbon black) screens out ultraviolet radiation. PE is classified into three types. Type I PE has a low density, is relatively soft and flexible, and has low heat resistance. Type II PE has a medium density, is slightly harder, is more resistant to higher temperatures, and has increased tensile strength. Type III PE is a high-density material offering maximum physical properties. Type III PE, owing to its toughness and superior physical properties, is the preferred piping material and will be used as the basis for the data presented in this text. (A newer material classification—ATSM D 3350—for PE piping materials will be covered in Chapter 4.) PE piping is used most frequently in gas distribution and water service markets.

Polypropylene (PP)—PP is a resin made by the polymerization of propylene. PP is the lightest weight piping material and has very good chemical resistance, even to many organic solvents. PP has two classifications. Type I is the most commonly used piping material; it is a homopolymer having better rigidity and strength but lower impact resistance than type II. Type II PP is composed of copolymers of propylene and ethylene or other olefins. Type II has improved toughness over type I material. PPs have similar additives to its base resin as polyethylene, with additional heat-stabilizer additives used to retard thermal aging; a flame retardant grade of PP is also available for drainage applications. PP is used in many applications involving more-corrosive and higher-temperature applications than PVC.

Polybutylene (PB)—PB is the most recent addition to the polyolefin family of plastics and is prepared by the polymerization of butylene. Piping material is designated as PB 2110, and normally contains 2% carbon black as an ultraviolet inhibitor. PB is well suited for piping, since it is flexible with long-term strength. Another significant feature of PB is its ability to retain its tensile strength better than most thermoplastics at higher temperatures. Its working temperature range is −10–200° F (−23–93° C) in a pressurized system. PB is used in hot and cold piping systems, and, owing to its abrasion resistance, is used successfully in slurry lines.

Acrylonitrile–Butadiene–Styrene (ABS)—ABS, a combination or blend of polymers, is used for piping materials in which the minimum butadiene content is 6%, the minimum acrylonitrile content is 15%, the minimum styrene content is 15%, and the maximum content of all other monomers is not more than 5% plus other additives. Similar to other plastics ABS is coded by short-term and long-term strength properties. Type I, grade 3 (ABS 1316) is an excellent resin for piping materials. ABS has toughness, strength, and stiffness, which account for its large usage in piping for drain, waste, and vent; sewers; well casings; and communication ducting. A specially formulated ABS material is one of the few plastic materials recommended for aboveground compressed air piping.

Styrene–Rubber (SR)—SR is made by combining styrene polymers with elastomers and other ingredients. SR piping compounds are not pressure rated and have lower impact strength and temperature resistance than ABS, but, are more rigid than most thermoplastics. SR piping is used exclusively for underground installations such as electrical conduit, sewer and drain pipe, low-head irrigation pipe, and foundation drains.

Polyvinylidine Fluoride (PVDF)—PVDF is produced by the polymerization of vinylidine with fluoride. PVDF has excellent chemical and solvent resistances, broad working temperature ranges, unexcelled resistance to weathering effects, and good strength. PVDF is more expensive than most other commonly used plastic piping materials and is used in highly corrosive environments (especially the halogens) where other piping materials prove inadequate.

Chlorinated Polyether (CPE)—CPE is a resin combining approximately 46% chlorine (by weight) with polymerized ether polymers. CPE was used almost exclusively for higher temperatures [up to 225° F (107° C)] and chemically aggressive fluids, but has been replaced by PVDF and other fluorocarbon piping materials. Hercules Inc. tradename for this material is Penton.

Polyvinylidene Chloride (PVC$_2$)—PVC$_2$ is a resin produced by the addition of chlorine to the polymerization of vinylidene monomers. Saran is the commonly used resin name. Most PVC$_2$ material is used as a lining in steel piping and vessels for chemically corrosive applications.

Cellulose Acetate Butyrate (CAB)—CAB is a plastic made by compounding a cellulose acetate butyrate ester with plasticizers and other additives. The ester is a derivative of cellulose (obtained from cotton or wood pulp). CAB was used mainly in gas- and oil-field applications, but, owing to its low strength and moderate resistance to chemical attack, it has been replaced by other plastic materials.

Polytetrafluoroethylene (PTFE), Perfluoroalkoxy (PFA), Fluorinated Ethylene Propylene (FEP) (Teflons)—PTFE is a polymer consisting of recurring tetrafluoroethylene monomer. PFA is a copolymer of TFE and perfluorinated vinyl ether. FEP is a copolymer of TFE and hexafluoropropylene. Few materials have "best" labels, but the Teflons claim three. They have the best chemical resistance, the lowest dielectric constant, and the lowest coefficient of friction of almost *any* solid material. Teflons are not used more widely because they are expensive and joining methods for them are not as convenient as for other materials. Most Teflon applications in piping systems are as liners, gasketing, and seals.

THERMOSETS

Glass-Reinforced Epoxy—Epoxy resins are made from diglycidyl ethers of bisphenol A, novolacs, or cycloaliphatics. The epoxies are then cured with hardeners and other curing agents to produce a cross-linked material that has physical and chemical properties of both the base resin and the curing agent. E-glass is usually the fiber-reinforcing agent with C-glass used in more severe chemical service as a reinforcement in the lining of pipe. Epoxies, especially those cured with aromatic amines, have excellent corrosive properties and can handle temperatures up to 300° F (149° C).

Glass-Reinforced Polyester—These polyester resins are normally bisphenol A–fumerates, chlorinated and isophthalic in chemical composition, and reinforced with E-glass in the pipe walls and C-glass in the liner and, in very corrosive environments, the external layer. Usually, the isophthalics are used in less severe corrosive services than the bisphenol A–fumerates. Polyesters are normally used in applications up to 200–225° F (93–107° C).

Glass-Reinforced Vinylester—Vinyl resins are styrene-diluted and free-radical-initiated in composition, having catalyst and promoter systems similar to the polyesters. The vinyl resins normally exhibit better chemical resistance properties than their related polyester resins with less brittleness. Vinyl esters are the newest of the three popular thermosetting resins and have made tremendous strides in industrial applications.

Glass-Reinforced Furan—These resins are derived principally from furfuraldehyde and fur-

furyl alcohol and have excellent resistance to solvents and will not support combustion. Furan handles applications to 300° F (149° C). Furan is not an easy material to process and is used only if epoxies, polyesters, or vinyl esters will not work satisfactorily.

FUTURE MATERIALS

The plastics industry as well as the end user have always desired that "magic" material with a tensile strength of 20,000 psi and temperature resistance of 300° F (149° C). In addition to these characteristics, the plastic material must be easily and inexpensively manufactured. Are there such materials? Perhaps there are! In the last decade tremendous strides have been made by organic research chemists in discovering "space age" materials that in the very near future may replace not only metal but several of the existing plastic piping systems. A few of these promising materials are:

Ryton® [polyphenylene sulfide (Phillips Chemical compound)] is a semicrystalline thermoplastic with tensile strength approaching 20,000 psi and temperature capabilities up to 350° F (177° C). Presently, some pump casings and valve bodies are being manufactured with this material.

PEEK® [polyetheretherketone (ICI compound)] is a semicrystalline thermoplastic with tensile strength (filled material) of more than 30,000 psi and temperature capabilities of 450–500° F (232–260° C). Compared to other materials it is relatively expensive ($22–27/lb), but its unique properties may afford it a special place in valve manufacturing.

PES [polyethersulfone (ICI compound)] is an amorphous thermoplastic with a high glass-transition temperature allowing it to retain most of its physical properties to temperatures as high as 350° F (177° C). PES is easily moldable and when filled can offer a tensile strength to 20,000 psi.

Ultem® [thermoplastic polyimide (GE compound)] is a polyimide that offers high strength (25,000 psi) at 300° F (149° C). It also has excellent chemical resistance and is less costly than PVDF to manufacture.

Xydar® [liquid crystal aromatic polyester (Dartco compound)] does not have a glass-transition point and always remains a crystal. Because of this characteristic, Xydar can handle 15,000–16,000 psi tensile strength at continuous use temperatures of over 400° F (204° C). It is easily moldable and has very good chemical resistance.

Vectra® [liquid crystal aromatic polyester (Celanese compound)] is similar to Xydar but can offer a tensile strength twice that of Xydar. It may exhibit some excess creep problems, but it is a promising material in the future manufacturing of fluid-handling products.

PHYSICAL CHARACTERISTICS

Table 1-4 is provided in order that the design engineer and user may have a handy reference guide to the capabilities of various plastic materials. (For information on commonly used plastic terms, consult the Glossary in Appendix A.) The actual characteristics of plastic pipe, fittings, and valves will not be exactly as shown on the chart, but rather these values will fall into a range depending on the additives to the basic resin. For the most part, the physical characteristics shown in the table are an averaging of the particular characteristics of the plastics normally used in piping compounds.

Engineering definitions of common expressions used in Table 1-4 follow:

Specific Gravity—The ratio of the density (mass per unit volume) of a material to the density of water at standard temperature.

Water Absorption—The percentage of water absorbed by a specimen immersed for a given period of time.

Tensile Strength—The pulling force necessary to break a specimen, divided by the cross-sectional area at the point of failure.

Modulus of Elasticity—The ratio of the stress to the elongation per inch due to this stress, in a material that deforms elastically.

Flexural Strength—The strength of a plastic material in bending. It is expressed as the tensile stress of the outermost fibers of a bent test sample at the instant of failure.

Izod Impact Strength—The resistance a notched

Table 1-4. Physical Properties of Commonly Used Plastic Piping Materials

Physical Properties[a]	Test Method (ASTM)	ABS (1210)	PVC (12454-B)	CPVC (23447-B)	PB (2110)	PE (355434C)	PPRO (Type 1)	PVDF	RTRP[b] (Epoxy)	RTRP[b] (Polyester)	RTRP[b] (Vinylester)
Color	—	Black Light Blue	Dark Gray White Blue Green	Beige Light Gray	Light Blue Black	Black Orange Tan	Black Natural White	Red Natural Black	Varies with manufacturer	Varies with manufacturer	Varies with manufacturer
Specific Gravity	D-792	1.04	1.38	1.55	0.92	0.95	0.91	1.76	1.90	1.60	1.90
Water Absorption [% in 24 hrs @ 73°F (23°C)]	D-570	0.3	0.05	0.05	0.01	0.01	0.02	0.05	0.03	0.03	0.03
Tensile Strength [psi @ 78°F (26°C)]	D-638 D-2105	4,500	7,200	8,400	3,800	3,300	5,000	6,000	35,000 hoop 9,000 axial	35,000 hoop 9,000 axial	35,000 hoop 9,000 axial
Modulus of Elasticity in Tension [psi @ 73°F (23°C) $\times 10^5$]	D-638 D-2105	3.0	4.2	4.2	3.5	1.2	1.7	2.1	14.0	14.0	14.0
Flexural Strength (psi)	D-790	10,000	14,500	15,350	3,000+	3,000	9,700	7,000	20,000+	20,000+	20,000+
Impact Strength (Izod) (ft-lb/in. notch)	D-256	6	0.7	3.0	7.5	7.0	2.1	3.8	25+	25+	25+
Coefficient of Thermal Expansion (in./in./°F $\times 10^5$)	D-696	5.0	3.0	3.8	7.2	7.8	5.5	7.9	0.6 hoop 1.4 axial	0.6 hoop 1.4 axial	0.6 hoop 1.4 axial
Thermal Conductivity (BTU/hr/ft²/°F/in.)	C-177	1.35	0.7	3.0	1.5	7.0	2.1	3.8	1.8	1.3	1.3
Heat Resistance at Continuous Drainage [°F (°C)]	—	180 (82)	150 (66)	210 (99)	210 (99)	160 (71)	180 (82)	280 (138)	300 (149)	230 (110)	250 (121)
Support Combustion	—	Yes	No	No	Yes	Yes	Yes	No	No	Yes	Yes

[a]The presented data are estimates. Each pipe manufacturer's resin may differ slightly in properties depending on the resin additives and manufacturing procedures.
[b]Reinforced thermosetting resins have varying ranges of physical characteristics depending on size, glass content, glass orientation, and other manufacturing processes. The data shown are for filament-wound pipe and are averaged values.

test specimen has to a sharp blow from a pendulum hammer.

Coefficient of Thermal Expansion—The fractional change in a length of a specimen due to a unit change in temperature.

Thermal Conductivity—The time rate of transferring heat by conduction through a material of a given thickness and area for a given temperature difference.

Heat Resistance—The maximum allowable temperature to which a piping system should be subjected.

Support Combustion—A specimen is ignited with a flame; the flame is then removed. If the specimen keeps burning after the flame has been removed, the plastic specimen is said to support combustion.

CHEMICAL RESISTANCE

Compared to other materials, plastics have excellent corrosion resistance. There is no galvanic or electrochemical attack owing to the nonconductance of plastics. Environmental attack is by direct chemical aggression, solvation (the absorption of chemicals into material causing softening and swelling), and/or stress (or strain) corrosion. Table 1-5 is a general guide of chemical resistance of plastic piping materials to grouping of chemicals. In Appendix B a more definitive table lists chemical-resistance data of plastics on over 600 chemicals. All representative chemical resistance data are based on the best information available, but still remain a guide. For marginal or critical applications, the incorporation of immersion testing using ASTM D-593 may prove helpful.

ABS and Teflons are not shown on the chemical-resistance tables. ABS is used mostly for drain, waste, and vent applications in home installations and is not normally subjected to chemically abrasive or very corrosive environments. Teflons are just the opposite. When no other material will handle a particular chemical, PTFE, PFA, or FEP is the preferred (or in many cases the only) material. Except for fluorine and related oxidizing compounds such as chlorine trifluoride, and violent reducing agents such as metallic sodium and other alkali metals, there are very few if any other materials that the Teflons cannot handle, in many cases up to 300–400° F (149–204° C). Premium grade resin data have been used to determine chemical resistance of the thermosetting material.

Table 1-5. Chemical Resistance of Plastic Piping Materials with Chemical Groups[a]

Chemical Groups	PVC	CPVC	PP	PVDF	PE	RTR Epoxy	RTR vinyl/ polyester
Inorganic							
Acids, dilute	R	R	R	R	R	R	R
Acids, concentrated	R	R	L	R	L	NR	R
Acids, oxidizing	NR	NR	NR	R	NR	NR	L
Alkalies	R	R	R	R	R	R	L
Acid gases	R	R	R	R	R	NR	L
Ammonia gases	R	R	R	R	R	R	R
Halogen gases	L	L	L	R	L	NR	R
Salts	R	R	R	R	R	R	R
Oxidizing salts	L	L	R	R	R	NR	R
Organic							
Acids	R	R	R	R	R	L	R
Acid, anhydrides	L	L	L	L	L	NR	NR
Alcohols	R	R	R	R	L	L	L
Esters/ketones/ ethers	NR	NR	L	L	L	L	L
Hydrocarbons— aliphatic	L	L	L	R	L	L	L
Hydrocarbons— aromatic	NR	NR	NR	R	NR	L	L
Hydrocarbons— halogenated	L	L	NR	L	NR	L	L
Natural gas	R	R	R	R	R	R	R
Synthetic gas	R	R	L	R	L	L	L
Oils	R	R	R	R	L	R	L

[a]This table is a general guide. For more specific details see Appendix B. R = recommended; L = limited usage; NR = not recommended.

However, owing to the many varying resin liners and different methods of construction of RTR pipe, it is wise to check chemical resistance with the particular RTRP manufacturer.

The chemical resistances of three elastomeric materials are shown in Appendix B. These materials were selected because of their popularity as gasket and O-ring material in flanging and valving of plastic piping products.

WEATHERABILITY*

Plastics are used in outdoor environments, and their weatherability, especially the thermoplastics's, has been extensively tested to learn the specific long-term effects of weather on pipe. The resistance of plastics to the environment has been evaluated in controlled laboratory studies that simulate the types of chemical and physical changes that occur with outdoor exposure. Such evaluations give highly satisfactory indications of the service life of a particular plastic subjected to

*Data adapted from Plastics Piping Institute publication TR-18, *Weatherability of Thermoplastic Piping.*

continuous exposure to the elements. Years of successful use proves conclusively that plastic pipe, used outdoors, gives excellent results.

FACTORS INFLUENCING WEATHERING

There are at least five factors that will influence weathering of any material, including plastics:

1. **Sunlight** absorbed by materials results in material degradation and possible accumulation of heat. Solar energy may be sufficient to cause breakdown of polymer resin and create changes in compounding ingredients. Temperature varies considerably with both the season and the location and can make quite large differences in a material's condition. Heat from solar radiation can increase the temperature of directly exposed materials as much as 60° F (33° C) higher than ambient. Extremes of temperature over extended periods of time can cause physical damage. In addition, chemical reaction rates increase exponentially as the temperature increases.

2. **Rain** and **humidity** are two contributors of moisture, with the latter having the greater overall effect. In general, humidity, a moist continuum in constant contact with a material, can contribute to hydrolysis leaching, etc. Rain provides a washing and impacting action.

3. **Wind** carries impurities that can contribute to weathering effects normally made by erosion. Similarly, the absence of wind can allow an accumulation of air contaminants, as in a smog, which could contribute to material weathering.

4. **Gases,** the nature and quantity of which vary widely, are usually present in industrial areas and can result in undesirable chemical action on some materials.

5. The **geographical location** is a factor. Lesser effects are produced when there are less hours of sunlight per year and when radiation is less intense. For example, a specific period of exposure is more detrimental in Arizona than in New Hampshire.

WEATHERING OF PLASTIC PIPE

As a result of experience and experimentation, thermoplastics and thermosets now include ultraviolet absorbers, antitoxins, and other additives, which prevent the degradation due to weathering of plastic pipe. In the present state of technology, these additives, which have resulted in such satisfactory outdoor performance, tend to make the pipe opaque. Generally, transparent and translucent plastic pipe, fittings, and sheet may not have as satisfactory a resistance to deterioration in outdoor environments.

The incorporation of carbon black into olefins greatly increases their weather resistance. The carbon black screens the olefins from damaging ultraviolet radiation. The resistance to aging, imparted by the carbon black, depends on its type, particle size, concentration, and degree of dispersion. A high concentration of carbon black, e.g., 2%, is particularly desirable.

Rigid PVC pipe has excellent weather resistance if it is made with suitable pigments and properly compounded. An air-conditioning system was installed in Orlando, FL, with rigid PVC pipe used outdoors. This pipe has been fully exposed on a roof for at least 20,000 hr of strong Florida sunshine with temperatures reaching 150° F (66° C). Results show its excellent resistance to weather deterioration; the pipe still continues to perform satisfactorily.

ABS pipe usually contains carbon black for protection from sunlight. Effects of ultraviolet radiation from the sun are reduced substantially in ABS pipe, permitting its use in many outdoor applications. The largest outdoor use is probably as vent pipes in drain, waste, and vent systems, which are exposed to all types of weather.

PP pipe installations involving prolonged exposure to weather have been documented and show very little weathering effects.

Thermosets, such as epoxies, polyesters, and vinylesters, can incorporate additives that give excellent outdoor weathering protection. Some product manufacturers recommend painting outdoor piping systems with an opaque material for the thermosets, but the weathering effects of a properly compounded thermoset material are minimal. Manufacturers usually have many years of experience and can determine what particular additives will give the best results outdoors. Always specify, when ordering material, that the usage is to be outdoors to make certain that the supplier has in stock the proper material to offer.

In summary, plastic pipe and fittings remain successfully in use outdoors after many years of operation, yet show minimal effects of weathering. Additional case histories of successful outdoor use of plastics usually may be obtained from any reputable resin or product manufacturer upon request.

COMBUSTION AND TOXICITY OF PLASTICS

In the United States there is a movement, at times verging on hysteria, to discredit plastic piping as a dangerous material. The culprits of this propagandized campaign are small but well-financed special-interest groups that are telling the public that plastic piping and conduit *causes* or *promotes* fires and that the fire's by-product, smoke, has poisonous gases that kill humans. Most of the clamor concerns PVC, which has secured over 55% of the U.S. conduit and drain, waste, and vent markets and as a material comprises close to 25% of the more than six billion feet of U.S. produced pipe. Fortunately, there have been many scientific and well-documented studies delving into the effects of fire involving plastics and other materials. The results and findings are: plastics in piping does not cause any more harm to human life than any other piping material. Some of the comments and conclusions of these studies follow.

Combustion—Wool, cotton, wood, paper, and plastics are organic materials that can be made to burn. Table 1-4 shows which plastic piping material support combustion and which do not. Supporting combustion means that, when an open flame ignites a material, that material will continue to burn even when the open flame is removed. Materials that do not support combustion are self-extinguishing when the open flame is removed. PVC, CPVC, PVDF, and RTRP epoxies do not support combustion.

Ignition—Most materials have a temperature of ignition at which the material will ignite, burn, and melt or evaporate. Plastic ignition temperatures are less than other piping materials, but in no condition can any plastic piping material flash or self-ignite until temperatures approach 650° F (343° C) or more depending on the material. Wood, paper, and cotton have ignition temperatures of 400–500° F (204–260° C) and support combustion. All plastic piping materials are joined with tools requiring *no open flame.* Other piping materials such as copper use an open flame for joints and have caused fires by igniting wood-framed

Table 1-6. Ignition Temperatures of Selected Plastic Piping and Other Materials[a]

Material	Ignition Temperature,[b] °F (°C)
Newspaper	445 (229)
Cotton	490 (254)
Douglas fir	500 (260)
Polyethylene	660 (349)
PVC	735 (391)
FRP polyester	750 (399)
ABS	780 (416)
PVDF	790 (421)
PTFE	986 (530)

[a]Adapted from Hilado's *Flammability Handbook for Plastics* and Tewarson's, *Physico-Chemical and Combustion/Pyrolysis Properties of Polymeric Materials* (see *Bibliography* for full source data).
[b]The lowest temperature is shown when there is a difference between flash and self-ignition temperatures.

structures. Table 1-6 lists the ignition temperatures of various materials.

Flame Spread—When forced to burn, many plastic piping materials's rate of burning is much lower than that of wood: The lower the rate of burning, the slower the oxygen consumption, the less heat released, and the slower the production of carbon monoxide. Lessening flame spread is critical to limiting the size of fires. Table 1-7 shows the limiting oxygen indices (LOI) of polymers (a low LOI value indicates high flammability).

Flaming Drops—Flaming molten drops are by-products of some materials. This characteristic can cause fires to spread more rapidly. Vinyls (PVC and CPVC) *do not* produce flaming drops, but when forced to burn, produce a harmless carbonaceous char.

Electrical Conductance—Plastic piping does not conduct electricity and does not rust or corrode. Metal piping, on the other hand, is a conductor of electricity and does corrode from electrolytic or galvanic attack, which can cause shorting. Short-

Table 1-7. Limiting Oxygen Indices (LOI) of Polymers[a]

Materials	LOI	
Cotton	16–17	
PE	17	
PP	18	
PS	18	Material burns by itself
ABS	19	
Fir	21.5	
Red oak	23	Material may burn upward[b]
Nylon	23	
PVC	40–49	
PVDF	44	
CPVC	45–60	Material will not burn by itself
PTFE	95	

[a]Adapted from B.F. Goodrich Bulletin No. 11, *Fire Characteristics of Rigid Vinyl* (See *Bibliography* for full source data).
[b]Burning may continue if a vertical sample is ignited at the bottom; however, burning may not occur if ignition is attempted at the top of the sample.

circuiting and electrical arcing can occur with metal piping and conduit and lead to tragic results. In 1980 at the MGM Grand Hotel in Las Vegas, dozens of deaths were attributed to fire caused by the arcing of metal conduit in the hotel.

Flammability Rating—Underwriters Laboratory has a standard for classifying polymeric materials's degree of flammability. Flammability ratings are rated 94 V-0 (the most resistant to burning) to 94 V-2 (the least resistant). To achieve a V-0 rating, the tested material must conform to the following requirements:

A. The tested material must not exhibit flaming combustion for more than 10 sec after flame is removed.

B. The tested material may not burn to the sample holding clamp after two applications of flame.

C. No flaming drops can drip and ignite cotton placed 12 in. below the test specimen.

D. No glowing combustion for more than 30 sec after a second application of flame.

PVC, CPVC, PVDF, and other fluorocarbons have 94 V-0 ratings.

Toxicity—Most plastic piping used in residential and industrial applications is approved by the National Sanitary Foundation (NSF), an independent agency, whose duties are to inspect and certify that manufacturers of plastic piping and piping components adhere to strict standards that ensure no possible danger to humans from drinking fluids being transported by plastic piping. The Food and Drug Administration also certifies many of the plastic piping systems used in transporting products in the food and beverage industry. In 40 years of manufacturing in this country, there has never been a reported case of injury or death caused by contamination of potable water or other food products being transported by plastic piping.

As was previously mentioned, there has been much recent concern about the toxic risks associated with plastic piping (mostly PVC) that is forced to burn. The facts are that plastic piping, like any other organic material, gives off combustion products when burning and, depending on the *mass* of plastic in combustion, toxic gases are released. Knowing that an organic material, when burning, gives off toxic gases, the next log-

ical step is to examine where the human hazard lies in fires. In almost every documented fire fatality studied, carbon monoxide was the primary toxicant. In most homes the presence of wood structures and furniture, organic carpeting/clothing/bedding, and other organic fixtures far outweigh in mass any home plastic piping system. Consequently, any additional toxic fumes added to a fire by plastic piping would be minuscule compared to the toxic gases given off by other home products. With dozens of studies and detailed investigative laboratory tests performed, there has never been any documented proof that plastic piping caused an additional risk of combustion toxicity compared to any other organic material.

Fire safety is an important subject and should be examined in a rational and nondiscriminatory manner. It should be noted that between 1940 and the present fire deaths on a per capita basis have decreased 40% as the use of plastic piping in home construction has increased greatly. Recently, Underwriters Laboratory and FM with the cooperation and guidance of the National Fire Protection Association have approved two plastic piping materials, CPVC and PB, for fire sprinkler systems in home and commercial uses.

MICRO- AND MACROBIOLOGICAL EFFECTS*

The micro- and macrobiological effects on plastics are of great concern. Fungi, a severe problem during World War II, especially in tropical and subtropical climates, caused the deterioration of electrical equipment. Malfunctions of communication equipment owing to the attack of living organisms is a continuing problem. Rodents can cause damage to underground power and communication cables, and, in most climates and soils, termites are responsible for considerable damage to organic materials.

There has been some attempt, particularly in the thermoplastic field, to gather information regarding the resistance of plastic piping materials to micro- and macrobiological degradation.

---

*Data adapted from Plastics Piping Institute Report TR-11, *Resistance of Thermoplastic Piping Materials to Micro- and Macro-Biological Attack.*

FUNGI

Fungi refers to a family of heterotrophic plant life consisting of mold, mildews, mushrooms, etc. They are lacking in chlorophyll and are unable to derive energy directly from sunlight. Instead, they derive their energy from organic materials such as carbohydrates, which are a particularly good nutrient for fungi.

Fungi normally thrive in warm, humid atmospheres and are abundant in tropical areas. Temperatures of 78–85° F (26–29° C) and relative humidities of 85–100% are most favorable for fungi. From the published literature, it is apparent that the growth of fungi on plastics is not due to the nutrient value of the polymer resin, but rather to lower-molecular-weight additives such as lubricants, stabilizers, and plasticizers. Thermoplastic materials used for the manufacturing of pipe contain little, if any, nonpolymeric material and, consequently, have a high degree of resistance to attack by fungi. Fungi may grow on the pipe surfaces by feeding on such nutrients as fly ash, which may settle on the surface. Similar growths are commonly observed on concrete and glass, which, like plastics, serve merely as a physical support for the life forms.

BACTERIA

Bacteria, in general, require a more moist environment than fungi. Some bacteria require the presence of oxygen to sustain life, while others are anaerobic and grow only when there is no oxygen. Since bacteria are encountered in nearly all areas where water is present, it should be expected that, when pipe is installed in wet areas, the pipe will come into contact with some forms of bacteria.

Laboratory tests show that situations between plastics and bacteria are similar to plastics and fungi, i.e., no nutrients are present in the plastic pipe composition. There have been some instances where rigid PVC pipe has been discolored when buried in mud under seawater, the discoloration being due to hydrogen sulfide produced by anaerobic sulfate-reducing bacteria. However, the mechanical properties of the discolored material were unchanged.

TERMITES

Termites are found worldwide and are a cause of extensive damage to wood. There have been many tests of plastic pipe buried in a termite-infested soil and periodically dug up and examined. PE, PVC, and PP are just some of the pipes that have been tested. When the soil was removed, these pipes were found to be free of termite infestation.

There have been termite attacks reported on plastic film and wire and cable insulation. However, the plastics attacked were softer and highly plasticized, containing additives for high flexibility, in contrast to rigid plastic pipe.

RODENTS

All materials, with the exception of the hardest metals, such as concrete, are capable of being gnawed by rodents. There have been instances where plastic pipe has been damaged by rodents, but these are of random nature, where they were trying to reach the water running through the pipe. Rodents have caused greatest damage to *plasticized pipe*, but with the discontinuance of this type of plastic piping, rodent infiltration has virtually been eliminated.

There are few data available on rodent, termite, and fungi attack on thermoset piping. To date, there are no documented reports of any micro- or macrobiological attacks on fiberglass-reinforced pipe.

2
General Design

INTRODUCTION

To use plastics properly in piping systems, it is important to have an adequate knowledge of piping design and to be aware of several of the unique properties of plastics.

The four most important factors to be considered in designing piping systems in plastics, especially thermoplastics, are

1. Temperature limitations

2. Pressure limitations

3. Temperature–pressure relationship

4. Expansion contraction.

PRESSURE RELATIONSHIPS IN PLASTIC PIPE

DETERMINATION OF OPERATING PRESSURE

Hoop or circumferential stress is the single largest stress present in any piping system under pressure. Consequently, it is the governing factor in determining the pressure that a pipe section can withstand. It is expressed by the equation:

$$S = \frac{P(D_o - t)}{2t} \qquad (2.1)$$

where

$$
\begin{aligned}
S &= \text{hoop stress, psi} \\
P &= \text{internal pressure, psi} \\
D_o &= \text{outside pipe diameter, in.} \\
t &= \text{minimum wall thickness, in.}
\end{aligned}
$$

The Plastic Pipe Institute, in conjunction with the American Society for Testing and Materials, has developed a standard method for determining the long-term hydrostatic strength of plastic pipe.

From these tests the S values in Eq. (2.1) are calculated, and a hydrostatic design basis (HDB) is obtained. This HDB is then multiplied by an appropriate service (design) factor to obtain the hydrostatic design stress.

Two general groups of conditions comprise the service (design) factor. The first group takes into account manufacturing and testing variables—specifically, normal variations in the material; manufacture; dimensions; good handling techniques; and evaluation procedures in the testing methods—to arrive at the HDB. Normal variations in this group are usually within 1–10%.

The second group considers the application or use—specifically, installation; environment (both inside and outside the piping); temperature; hazard involved; life expectancy desired; and degree of reliability selected.

For a piping system handling water at 73° F (23° C), the service (design) factor recommended by the Plastics Pipe Institute is 0.50. Consequently, the hydrostatic design stress for such a system would be the HDB multiplied by 0.50.

Other service factors are given in Table 2-1.

The problems of developing service factors are numerous, since there are variations among the different plastic materials as well as differences due to the corrodent being handled. All of these differences are influenced by temperature as well.

Temperature effects have been studied and correction factors have been developed by piping manufacturers. These factors can be used in con-

Table 2-1. Service Factors

Service (Design) Factor	Temperature	Medium	Special Circumstances
0.40	37.8° C (100° F)	Water	Five specific polyethylene, and two specific polyvinyl chloride, pipe compounds
0.20	60° C (140° F)	Water	For CPVC 4116 and CPVC 4120
0.125	82.2° C (180° F)	Water	For CPVC 4116 and CPVC 4120
0.32	37.8° C (100° F)	Natural Gas	No allowance is made for excessive amounts of odorants, antifreeze, and aromatic hydrocarbons

junction with the allowable operating pressures at 73° F (23° C).

Once the service (design) factor has been established, Eq. (2.2), derived from Eq. (2.1), is used to determine the maximum allowable operating pressure:

$$P_w = \frac{2S_w t}{D_o - t} \qquad (2.2)$$

where

P_w = pressure rating, psi
S_w = hydrostatic design stress, psi (hydrostatic design basis × service [design] factor)
t = minimum wall-thickness, in.
D_o = outside diameter of pipe, in.

The pressure rating represents the maximum pressure allowable in an operating system including any surge pressures. Maximum operating pressures of commonly used plastic pipe are shown in Table 2-2A.

TEMPERATURE CORRECTIONS

Most pressure ratings for thermoplastic pipe are determined in a water environment of 73.4° F (23° C) plus tolerances of 3.6° F (2° C). As the temperature of the environment increases, the thermoplastic pipe becomes more ductile. This effect causes an increase in the impact strength and a decrease in the tensile strength of the pipe. Because of this phenomenon, the pressure rating of thermoplastic pipe must be decreased to allow for safe operation. Thermosets are not subjected to temperature correction factors. Temperature correction factors for plastic pipe are shown in Table 2-2B.

THERMAL EXPANSION AND CONTRACTION OF PLASTIC PIPE

Most plastic materials, especially thermoplastics, exhibit a relatively high coefficient of thermal expansion. Thermoplastics have a coefficient

Table 2-2A. Maximum Operating Pressures of Commonly Used Plastic Pipe [psi at 75° F (24° C)][a,b,c]

Nominal Pipe Size, in. (IPS)	PVC		CPVC	Polypropylene	Polyethylene UHMW (PE 34)	PVDF	RTRP
	Schedule 40 Socket	Schedule 80 Socket	Schedule 80 Socket	Schedule 80 Socket	Schedule 80 Socket or Fusion	Schedule 80 Socket	Filament-Wound (Typical) Socket
1/2	600	850	850	410	Not Available	480	150
3/4	480	690	690	330	176	400	150
1	450	630	630	310	164	380	150
1 1/4	370	520	520	250	134	Not Available	150
1 1/2	330	471	471	230	121	300	150
2	280	400	400	200	101	260	150
2 1/2	300	425	425	185	Not Available	Not Available	150
3	260	375	375	185	96	235	150
4	220	324	324	160	81	210	150
6	180	280	280	140	64	Not Available	150
8	160	250	Not Available	130	57	Not Available	125
10	140	230	Not Available	Not Available	51	Not Available	100
12	130	230	Not Available	Not Available	48	Not Available	100

[a]Threading of Schedule 80 pipe reduces the operating pressure by one-half. Schedule 40 pipe is not recommended for threading. Threaded polypropylene or polyethylene pipe should *not* be used in pressure applications.

[b]For pressure-rated thermoplastic pipe, the pipe has a constant stated pressure at 75° F. RTRP has a constant stated pressure up to a given maximum temperature.

[c]RTRP can have pressures vary depending on wall thickness.

Table 2-2B. Temperature Correction Factors of Commonly Used Plastic Pipe[a]

Operating Temperature °F (°C)	PVC	CPVC	Poly-propylene	Poly-ethylene	PVDF
75 (24)	1.00	1.00	1.00	1.00	1.00
80 (27)	0.90	0.96	0.95	0.95	0.95
90 (32)	0.75	0.92	0.90	0.88	0.90
100 (38)	0.62	0.85	0.85	0.82	0.85
110 (43)	0.50	0.77	0.80	0.76	0.80
115 (46)	0.45	0.74	0.78	0.72	0.78
120 (49)	0.40	0.70	0.75	0.69	0.76
125 (52)	0.35	0.66	0.73	0.66	0.73
130 (54)	0.30	0.62	0.70	0.63	0.71
140 (60)	.0.22	0.55	0.65	[b]	0.67
150 (66)	0.15	0.47	0.58	[b]	0.63
160 (71)	[b]	0.40	0.50	[b]	0.58
170 (77)	[b]	0.32	0.26	[b]	0.54
180 (82)	[b]	0.25	[b]	[b]	0.50
200 (93)	[b]	0.18	[b]	[b]	0.42
210 (99)	[b]	0.15	[b]	[b]	0.35
220 (104)	[b]	[b]	[b]	[b]	0.30
240 (116)	[b]	[b]	[b]	[b]	0.25

[a]To determine maximum operating pressures at a given temperature, multiply the maximum operation pressure by the correction factor.

Example. What is the maximum operating pressure of 1½" CPVC pipe Schedule 80, at 120° F (49° C)?

Solution: Maximum operating pressure of 1½" CPVC pipe = 471 psi. Temperature correction factor of CPVC @ 120° F·= 0.70. Therefore, the maximum operating pressure is 471 × .70 = 329.7 psi.

[b]Not recommended.

of axial expansion five to six times that of steel. For this reason, it is important to consider thermal elongation when designing piping systems. Thermosets [fiberglass reinforced plastics (FRP)] have a coefficient of expansion twice that of steel in the axial direction and about the same as steel in a hoop direction; hence, thermal expansion in thermosets is less critical in design than in thermoplastics. The main thing to remember in FRP material is that there exists a very low modulus of elasticity for FRP, which more than compensates for a higher coefficient of thermal expansion.

Generally, when designing with thermoplastics where runs exceed 100 ft, if the total temperature change exceeds 30° F (17° C), provision should be made to compensate for thermal expansion or contraction. The fact that the installation temperature, as well as the ambient temperature changes, has a bearing on the coefficient of thermal expansion must be kept in mind. "A piping system may be designed to handle liquids at ambient temperatures, but in an outside installation the ambient temperature change between winter and summer could very easily exceed 30° F. If threaded connections are used, caution should be exercised in the design of the expansion loops since threaded connections are more vulnerable to failure by

bending stresses than are solvent welded joints. It would be better to fabricate an offset expansion loop using solvent welded fittings regardless of what type of joint is used for the rest of the system."[1] To calculate the expansion or contraction in plastic pipe, use the following formula:

$$\Delta L = y \frac{T - F}{10} \times \frac{L}{100} \qquad (2.3)$$

where

ΔL = expansion of pipe, in.
y = constant factor expressing inches of expansion per 10° F temperature change per 100 ft of pipe
T = maximum temperature (°F)
F = minimum temperature (°F)
L = length of pipe run (ft)

Example: How much expansion can be expected of 300 ft of PVC pipe installed at 50° F and operating at 125° F?

Solution: From the accompanying table, y = 0.360, therefore

$$\Delta L = (0.360) \frac{125 - 50}{10} \times \frac{300}{100}$$

$$= 0.360 \times \frac{75}{10} \times \frac{300}{100}$$

$$= 8.1 \text{ in.}$$

There are several methods of handling expansion in plastic piping. The two major methods are using plastic-piston expansion joints and fabricated offsets or expansion loops. Bellows expansion joints have been used, but can present problems if not applied properly.

Value of y for Specific Plastics

Material	y Factor (in./10° F/100 ft)
PVC, Type 1	0.360
CPVC	0.456
Polypropylene	0.600
Polyethylene, UHMW	1.000
PVDF	0.948

PLASTIC EXPANSION JOINTS

Plastic-piston expansion joints (see Fig. 2-1) are available in PVC and CPVC from ½ up to 12-in. diameters with travel lengths of 6, 12, and 24 in.

[1]P. A. Schweitzer, *Handbook of Corrosion Resistant Piping*, Industrial Press, New York, 1969.

Fig. 2-1. Piston-type expansion joint. (Courtesy of Chemtrol Div. of Nibco.)

The expansion joints are made of two tubes, one telescoping inside the other, with an O-ring seal or packing. The outer tube must be firmly anchored, while the inner tube is permitted to move freely as the pipe expands or contracts. Alignment of the joints is critical, since binding will result if the pipe is misaligned and does not move in the same plane as the joint.

Guides should be installed approximately 1 ft from both ends of each expansion joint. In piping runs exceeding 150 ft, the pipe must be anchored at each change of direction so that the pipe movement will be directed squarely into the expansion joint. Joints are installed with the piston partially extended, depending on the ambient temperature. The correct piston setting is determined as follows:

$$O_P = \frac{T - A}{T - F} \times L_J \qquad (2.4)$$

where

O_P = piston "out" position (in.)
T = maximum operating temperature of pipe (°F)
A = ambient temperature at time of installation (°F)
F = minimum operating temperature of pipe (°F)
L_J = maximum required expansion length of joint (in.)

An example of sizing and setting an expansion joint follows:

Example: A straight run of 215 ft of PVC Type I pipe will operate at temperatures between 75 and 135° F. Size of joint required:

From Equation (2.3):

$$L_J \Delta L_{\max} = 0.360 \times \frac{135 - 75}{10} \times \frac{215}{100} = 4.6 \text{ in.}$$

A 6 in. joint will be adequate. Ambient temperature at time of installation is 85° F. Piston "out" position from Eq. (2.4):

$$O_P = \frac{135 - 85}{135 - 75} \times 6 = 5 \text{ in.}$$

Extend piston out 5 in. when installing.

EXPANSION LOOPS

When expansion cannot be accommodated by regular dimensional changes, there is a common way, if space allows, to use an offset expansion loop. The loop uses elbows and straight pipe joined by solvent cementing. The developed length of the offset, or loop, can be calculated by the following formula:

$$L_o = 3.5 \sqrt{D_o \times L_J} \qquad (2.5)$$

where

L_o = developed length of pipe in the expansion or offset (ft)
D_o = outside diameter of piping (in.)
L_J = expansion or contraction to be accommodated (in.)

The thermoplastic piping systems normally have a relatively low modulus of elasticity and can be designed to restrain the expansion and absorb it as an internal compression. The stress caused by expansion that is restrained below the elastic limit in a straight length of pipe is calculated by the equation:

$$S_c = Ee \qquad (2.6)$$

where

S_c = compressive stress in pipe wall cross section (psi)
E = Young's modulus of elasticity (psi)
e = strain or,

$$e = \frac{\text{change in length}}{\text{original length}}$$

In Fig. 2-2 several methods of handling expansion and contraction in thermoplastic piping are shown. Be certain that *no* restraints or supports are placed on the leg length of the offset or loop.

With proper guides and anchors, the piping system can be evaluated for stresses to insure against failure by buckling. Additional information on guides and anchors will be given later, in the section on "Aboveground Installation."

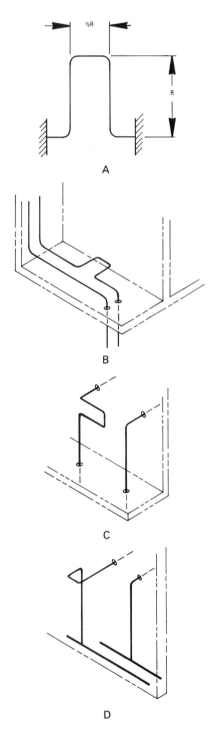

Fig. 2-2. Several methods of handling expansion and contraction in thermoplastic piping.

BELLOWS-TYPE EXPANSION JOINTS

The bellows-type fitting is normally available only in TFE or FEP for corrosion resistance, operating pressures, and operating temperatures comparable at least to the plastic piping systems being used, but these fittings should be checked for compatibility. The chief reason for using this type of expensive expansion joint is to handle movements other than axial and for possible use in shorter runs of pipe where installation space is reduced and expansion is less than 2 in.

Bellows-type joints are available in flexible couplings, expansion joints, and bellows. Regardless of the configuration, the connecting flanges are identical.

Care should be taken in the selection of the configuration to be used. As the number of convolutions in the joint increases, the allowable operating pressure and temperature decrease. When using expansion joints of this type, vacuum service limitations may be critical. The installation of bellows joints should be checked with the particular manufacturer and care should be taken to follow directions carefully, since excessive torques, pressures, and stresses can be applied to plastic pipe if the joints are not installed correctly. The maximum travel permitted in each bellows-type expansion joint is 2 in.

THERMALLY INDUCED STRESSES

Stress in piping systems exists for all piping materials. It is pertinent to examine the major differences between metal and plastic piping when calculating stresses caused by changes in temperature.

Thermally induced stress (S) is equal to a force or thrust (F) in pounds divided by the cross-sectional area of pipe (A_p) in in.:

$$S = F/A_p$$

This formula may be adapted to include the modulus of elasticity (E), coefficient of thermal expansion (C_T), and change in temperature (ΔT):

$$S = E \times C_T \times \Delta T$$

Using this formula and the values listed in Table

1-4, we can calculate stresses in both metal and plastic piping systems. Table 2-3 tabulates the stresses induced in piping systems with a $\Delta T = 70°$ F ($39°$ C). The stress values in Table 2-3 are calculated at $75°$ F ($24°$ C). For temperatures above $75°$ F ($24°$ C), the stresses will be similar in metals, but actually decrease in plastic piping owing to the decrease in the modulus of elasticity (E).

As shown in Table 2-3, thermally induced stresses are many times higher in metal than in plastic piping. The high stresses developed is a major reason that broken pipe supports, wall damage, and bent pipe anchors occur in poorly installed or designed metal piping systems. With plastic piping systems, if stresses are not properly designed for, pipe buckling (not failure) usually occurs first and rarely destroys any other wall or other support structures.

WATER HAMMER

Whenever the rate of fluid flowing in a pipeline is altered, there is a change in fluid velocity, causing surge. The longer the line and the faster the liquid velocity, the greater the shock load from surge will be. Shock loads can be of sufficient force to burst any pipe, fitting, or valve. This phenomenon is called "water hammer."

Maximum pressure caused by water hammer or momentum pressure surges, may be calculated from the following formula:

$$p = v\left(\frac{SG - 1}{2}C + C\right)$$

where

p = maximum surge pressure, psi
v = fluid velocity, ft/sec (fps)
C = surge wave constant for water at $73°$ F ($23°$ C)
SG = specific gravity of liquid

Example 1 (Courtesy of Chemtrol Div. of Nibco):

A 2 in. PVC schedule 80 pipe carries a fluid with a specific gravity of 1.2 at a rate of 30 gpm ($v = 3.35$ fps) and at a line pressure of 160 psi. What would the surge pressure be if a valve were suddenly closed?

From Table 2-4,

$$C = 24.2$$

therefore,

$$p = (3.35)\left(\frac{1.2 - 1}{2}24.2 + 24.2\right)$$

$$p = (3.35)(26.6) = 90 \text{ psi, about}$$

Total line pressure = 90 + 160 = 250 psi

Schedule 80 2 in. PVC has a pressure rating of 400 psi at room temperature. Therefore, 2 in. schedule 80 PVC pipe is acceptable for this application.

Table 2-4 lists the surge wave constants for plastic piping materials. When calculating the surge pressure, the static line pressure must be added to the surge pressure to obtain the maximum (total) system operating pressure. The maximum system operating pressure should not exceed 1.5 times the recommended working pressure of the system. (In the preceding example, the static

Table 2-3. Thermally Induced Stresses in Metal and Plastic Piping Materials

Material	E (psi × 10⁶)	C (in./in./°F × 10⁶)	Stress (psi) @ $\Delta T = 70°$ F ($39°$ C)
Metal			
Aluminum	10.0	12.4	8,680
Brass	15.0	10.4	10,920
Carbon steel	30.0	6.3	13,230
Stainless steel	28.0	5.7	11,172
Plastics			
ABS	0.30	50.0	1,050
CPVC	0.42	38.0	1,117
PP	0.17	55.0	655
PVC	0.42	30.0	882
PVDF	0.21	79.0	1,161
RTRP	1.40	14.0	1,372

Table 2-4. Surge Wave Constant

Pipe Size (in.)	PVC Sch. 40	PVC Sch. 80	CPVC Sch. 80	Polypropylene Sch. 80	PVDF Sch. 80
1/4	31.3	34.7	37.3	—	—
3/8	29.3	32.7	34.7	—	—
1/2	28.7	31.7	33.7	25.9	28.3
3/4	26.3	29.8	31.6	23.1	25.2
1	25.7	29.2	30.7	21.7	24.0
1 1/4	23.2	27.0	28.6	19.8	—
1 1/2	22.0	25.8	27.3	18.8	20.6
2	20.2	24.2	25.3	17.3	19.0
2 1/2	21.1	24.7	26.0	—	—
3	19.5	23.2	24.5	16.6	17.3
4	17.8	21.8	22.9	15.4	16.2
6	15.7	20.2	21.3	14.1	—
8	14.8	18.8	19.8	13.2	—
10	14.0	18.3	19.3	12.1	—
12	13.7	18.0	19.2	11.9	—
14	13.4	17.9	19.2	—	—

Adapted from Chemtrol Div. of Nibco Data.

line pressure of 160 psi was added to the calculated surge pressure of 90 psi to obtain the total line pressure of 250 psi.)

The principal causes for surge are the closing and opening of a valve, the starting and stopping of a pump, and the movement of pockets of air accumulated in the pipeline. In order to reduce the chances of damaging surge from occurring, it is recommended that the following steps be considered when designing, installing, and operating a pipeline of either plastic or metal:

1. In no case shall the actual maximum operating pressure plus the surge in any section of the pipeline exceed the specified pressure rating of the plastic pipe.

2. Keep the velocities down at all times. As a rule of thumb, the velocity should not exceed 5 fps, preferably not more than 3–4 fps. Velocity at start-up is especially important and should not be more than 1 fps during filling and afterward, until it is certain that the air has been flushed out and the pressure has been brought up.

3. Do not use a multiple-step reduction such as a 12 × 2-in. Tee. If a reduction of more than four pipe sizes is required, fabricate it with a smooth reduction, not with multistep bushings. This type of bushing could create excessive pressure buildup causing damage to the piping system.

4. Eliminate as much air as possible at start-up; then, at shutdown or emergency shutoff, allow air to re-enter so as to prevent possible damage from "column separation."

5. Prevent air from accumulating while the line is operating and delivering. Avoid air being drawn in by the pump.

6. Provide all necessary protective equipment for the system and any of its parts as conditions may determine.

7. Whenever a system is to be started up or restarted, the pump start-up procedure should include the following:
    A. Close pump discharge valve.
    B. Open any air exhaust valves or vents at the end of the line.
    C. Start the pump.
    D. Start to open the discharge slowly and allow the valve to fill.
    E. Provide for adequate air relief.
    F. As soon as the line is filled with water, continue to flush the water through at the rate of approximately 1 foot per second while the pressure continues to rise.
    G. When the increasing pressure begins to level off, continue flushing for approximately 15 min until little or no air exhaust occurs.
    H. Pressure is then slowly increased to the operating pressure.

Equipment used to protect piping systems are pressure relief valves, manually operated gate valves, control closing check valves, shock absorbers, surge arrestors, vacuum air relief valves; but the most important aspect in designing the pipeline is to realize that water hammer can occur and to minimize this effect as much as possible.

COLLAPSE RATING AND LOADING

Applications such as buried pipe or vacuum lines exist in which the external pressure on the pipe may exceed the pipe internal pressure. If this condition exists, it is imperative that the allowable collapse loading of the thermoplastic pipe not be exceeded. In other words, the collapse loading of the pipe is the difference between external and internal pipe pressure. Therefore, pipe with a 200 psi internal pressure can withstand 100 psi more of external pressure than a pipe with 100 psi internal pressure.

The collapse ratings of thermoplastic piping are shown in Table 2-5. These ratings are calculated using the piping material's compressive strength, modulus of elasticity, outside and inside pipe diameter, wall thickness, out of roundness factor, and Poisson's ratio. As is shown in the table, thermoplastic piping is quite suitable for vacuum applications. Experience shows that PVC systems can be evacuated to pressures as low as 5 microns with continuous pumping. When working with vacuum installation, the following conversion factors may be used:

1 standard atmosphere = 14.676 psia
1 inch of mercury = 0.4914 psi
1 inch of mercury = 254 mm of mercury
1 mm of mercury = 1000 microns

Table 2-5. Long-Term Collapse Ratings[a] of Thermoplastic Piping (psi)[b]

Pipe Size (in.)	PVC Sch 40	PVC Sch 80	CPVC Sch 80	PPRO Sch 80	PVDF Sch 80
$^1/_2$	450	575	575	335	975
$^3/_4$	285	499	499	295	525
1	245	469	469	275	400
$^1/_4$	160	340	340	205	225
$1^1/_2$	120	270	270	140	155
2	75	190	190	95	105
$2^1/_2$	100	220	220	—	—
3	70	155	155	75	85
4	45	115	115	50	55
6	25	80	80	30	35
8	16	50	50	20	25
10	12	43	43	18	20
12	9	39	39	17	19

[a]Long-term collapse ratings are approximately one-third the value of short-term ratings.

[b]Pressures shown are for 73° F (23° C). For higher temperatures, use the temperature correction factors shown in Table 2B.

When using thermoplastics in vacuum services, always use solvent cement joints with vinyls and heat-fusion joints with polyolefins and PVDF.

FRICTION-LOSS CHARACTERISTICS OF WATER THROUGH PLASTIC PIPE

One of the major advantages of plastic pipe is its smooth inside surface area, which minimizes the amount of friction loss compared to other materials.

The C factor in the Williams and Hazen formula (see nomograph, Fig. 2-3) is 150 for plastics, and less for metallic pipe (see Table 2-6 for other C factors). Essentially this means that in plastic systems there is a possibility of using a smaller diameter pipe while obtaining the same or lower friction losses than with other materials.

The nomograph, Fig. 2-3, shows the friction-loss characteristics of plastic pipe and the example and directions given in the nomograph demonstrate how to use it properly.

HOW TO USE THE NOMOGRAPH (FIG. 2-3) TO ESTIMATE FRICTION LOSS IN PLASTIC PIPE

The values of this graph are based on the Williams and Hazen formula (based on full cross-sectional flow in pipe):

$$f = 0.2083 \left(\frac{100}{C} \right)^{1.852} \times \frac{g^{1.852}}{d^{4.8655}}$$

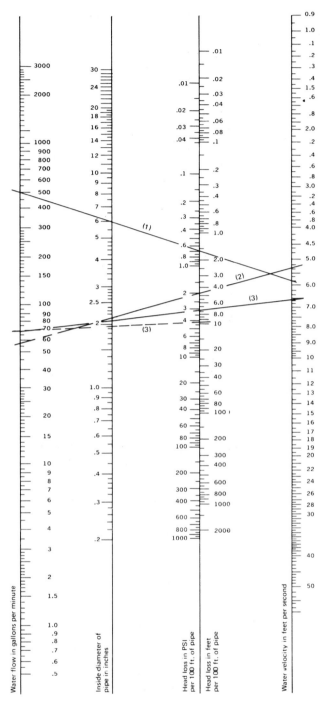

Fig. 2-3. Hydraulic design of water distribution mains or friction loss in rigid plastic pipe.

where

f = friction head in feet of water per 100 ft

d = inside diameter of pipe in in.

g = flowing gallons per min

Table 2-6. *C* Values for Various Piping Materials

Constant (*C*)	Type of Pipe
150	Plastic piping
	New steel pipe or tubing
140	Glass tubing
	Copper tubing
	Ordinary brass pipe
	Cast iron—new
	Cast iron—tar coated but new
130	Cast iron—fully cement lined
125	Steel pipe—old
	Wood stave pipe
	Concrete pipe
	New wrought iron pipe
120	Four- to six-year-old cast iron pipe
	Ten- to twelve-year-old cast iron pipe
	Vitrified pipe
	Spiral riveted steel, flow with lap
110	Galvanized steel
	Spiral riveted steel, flow against lap
	Thirteen- to twenty-year-old cast iron pipe
	Galvanized steel—over 5 years old
100	Cast iron—tar coated over 10 years old
90	Twenty-six to thirty-year-old cast iron pipe
60	Corrugated steel pipe

C = constant for inside roughness of the pipe (C = 150 for thermoplastic pipe)

Examples

(a) For a required service of 500 gpm in a pipe having an ID of about 6 in., what would be the head loss?

Using a straight edge, line up 500 in the first (left most) nomograph scale with 6 in the second scale (see line 1). This line intersects a head loss per 100 ft of about 0.7 psi or 1.6 ft in the third nomograph scale. Also note from the fourth scale that the flow velocity would be about 5.7 fps under these conditions.

(b) Similarly, if a straight edge were positioned as line 2 in the nomograph, a flow of 58 gpm, at a head loss of 2 psi per 100 ft would occur in a pipe having an approximate ID of 2.1 inches. Flow velocity would be approximately 5.3 fps.

(c) For the lines numbered 3, it can be seen that for a flow of 70 gpm, a change in inside pipe diameter from approximately 2 to 2.1 inches reduces the friction loss in the pipe from about 4 to 3 psi per 100 ft of pipe.

In Appendix C, columnized friction-loss tables, and various sizes and schedules of pipe are shown. This might be easier to use than the nomograph. Included in Appendix C is the estimated friction loss through fittings of particular size and geometry.

VALVE FRICTION LOSS

Most plastic valve manufacturers publish flow coefficients (C_v) for their valve products. These values are determined by laboratory testing and are defined as the flow rate through the valve required to produce a pressure drop of 1 psi. The following formula may be used to determine pressure drop through a valve for a given condition:

$$\Delta P = \frac{Q^2(SG)}{C_v^2}$$

where

ΔP = pressure drop across the valve, psi
Q = flow through the valve in gpm
SG = specific gravity of the liquid (water = 1.0)
C_v = flow coefficient

Example   Find the pressure drop across a $1\frac{1}{2}$ in. PVC ball check valve with a water flow rate of 50 gpm and a C_v of 100.

$$\Delta P = \frac{(50)^2 \times 1.0}{(100)^2}$$

$$\Delta P = \left(\frac{50}{100}\right)^2$$

$$\Delta P = 0.250 \text{ psi}$$

INSULATION OF PLASTICS

Plastic piping, when compared to metal pipe, has an unusually low thermal conductivity (K value). For example, steel's thermal conductivity is 314 and copper transmits heat even more readily, having a thermal conductivity of 2700 btu-in./ft²-°F-hr. With PVC and other plastics having a thermal conductance 1/300th of steel and 1/2700th of copper, the nonconductance of plastics allows the pipe not only to transmit the fluid but also to act as an insulating material. Therefore, when using plastic piping, there may be no need to insulate a particular piping system, whereas a similar metal piping system would require insulation to prevent sweating or to maintain a desired energy level.

To calculate heat loss or gain through plastic piping with no insulation, and knowing the inside and outside pipe wall temperatures, the following equation is used:

$$q = \frac{2K\pi L \Delta T}{D/d \text{ (in.)}}$$

where

q = heat gain or loss (Btu/hr)
K = thermal conductivity of the pipe (Btu-in./ft²-hr-°F)
ΔT = temperature difference of inside and outside pipe walls (°F)
L = unit length (in.)
D = outside diameter (in.)
d = inside diameter (in.)

Heat loss/gain calculations may also be used in more complex situations in which determination of gains and losses are subjected to variables of heat transfer coefficients, viscosity, fluid conductivity and fluid heat capacity. Also useful, especially when designing a piping system to perform to an engineer's desired energy requirements, are the following two equations. The first equation

$$q = \frac{\pi L \Delta T}{\dfrac{D/d \text{ (in.)}}{2K_p} + \dfrac{D_{ins}/D}{2K_{ins}} + \dfrac{1}{h_{ins} D_{ins}}}$$

where

K_p = mean thermal conductivity of pipe (Btu/hr-ft²-°F)
K_{ins} = mean thermal conductivity of insulation (Btu/hr-ft²-°F)
D_{ins} = outside diameter of pipe plus the insulation (in.)
h_{ins} = inside air contact coefficient between the insulation and the weather barrier (Btu/hr-ft²-°F)

determines the actual amount of heat gain when increased insulation thicknesses are used. The second equation determines the surface temperature of the insulation:

$$T_s = T_{air} - \frac{qR_s}{\pi}$$

where

T_s = surface temperature of insulation at actual thickness (°F)
q = actual heat gain based on actual insulation thickness (Btu/hr)
T_{air} = maximum possible ambient air temperature (°F)
R_s = surface resistance

Here,

$$R_s = \frac{1}{H D_{ins}}$$

where

H = outside heat transfer coefficient (Btu/hr-ft²-°F)
D_{ins} = OD of pipe plus insulation (in.)

These two equations form the basis for calculating heat loss values for plastic piping in still-air conditions.

Instances where insulation of plastic pipe should be considered are:

1. Cold applications in the range of 32–50° F (0–10° C) where there is a possibility of pipe sweating on a hot day. Depending on the application, pipe sweating might be damaging to the particular environment (especially in food preparations).

2. In some low-temperature processes where the change in temperature between the refrigeration medium and the process is small, the heat loss may be objectionable for economic reasons. A heat balance of individual processes in this type of application becomes a necessity, so that an engineer can judge the economics of insulation for a particular process operation.

3. Additional insulation for plastic pipe might be required to prevent freezing of liquids in outdoor installations.

OTHER ABOVEGROUND DESIGN CONSIDERATIONS

COLD WEATHER

Most plastic piping systems work quite well in temperatures below 0° F (−18° C). In fact, tensile strength increases in thermoplastic piping as temperature decreases. However, the impact resistance of most thermoplastics decreases with a decrease in temperature, and brittleness appears in these piping materials. As long as the piping material is undisturbed and not jeopardized by blows or bumping of objects and the piping is not dropped from truck beds (the pipe could actually shatter or splinter if the temperature was at freezing or below), there will be no deleterious effects to the plastic piping.

To protect plastic piping and pipe traps from freezing, do not use alcohol or petroleum distillates for antifreeze; instead, use only approved plastic pipe antifreeze or one of the following solutions:

1. 60 weight% of glycerin in water mixed at 73° F (23° C).

2. 22 weight% of magnesium chloride in water.

3. Strong solutions of common table salt.

Ultra-high-molecular-weight, high-density polyethylene (PE) and polybutylene (PB) are tough and ductile piping materials and may handle many cycles of freezing and thawing with no apparent damage to the pipe. For solvent cementing pipe in cold weather, see Chapter 3.

HOT ENVIRONMENTS

When pressure piping applications exceed temperatures of 300° F (149° C), the use of solid plastic piping systems is very limited. There are plastics today that are capable of handling temperatures above 400° F (204° C), but, except for lined metal piping, they are not commercially available in solid piping materials. Be certain to determine the installation's maximum external as well as internal temperatures when selecting the piping material. In some cases not designing for the piping external temperature could cause excessive sagging due to lack of pipe supports. Be certain that you label or design a plastic piping layout so that workmen cannot mistakenly fill it with very hot liquids or steam. If this possibility exists, you may want to use CPVC instead of PVC or RTRP epoxy instead of polyolefin piping. Finally, be certain that, when designing with plastics, you have taken into account the nature of thermoplastics. That is, when the temperature increases, the tensile strength and working pressure decrease. The necessary data to design for these conditions are listed in this chapter.

THREADING PLASTICS

From a structural design viewpoint, threaded plastic piping systems (especially thermoplastic systems) are one of the weakest joining systems available. The notch sensitivity of the plastic materials makes threaded areas the weakest point in the piping systems. This weakness is especially critical if there are any appreciable bending moments in the piping. In thermoplastic threaded systems, only schedule 80 or 120 pipe is recommended. The pressure ratings of these systems when used in threaded pipe are decreased 50%. In polyolefin, threaded piping systems (PE, PB, or PP) are not recommended above 20 psi working pressures. Solvent cemented and fused joining systems are much more reliable and offer complete piping system integrity.

PLASTICS AND COMPRESSED AIR

Compressed air is a potential source of trouble in most plastic piping systems. No manufacturer recommends plastic piping for aboveground compressed air systems except one; this particular supplier has a specially formulated ABS compound designed for compressed air. Compressed air in some piping materials such as PVC and CPVC, if at high enough pressures and volume, could cause shattering of the piping system and physical and personal harm.

The best way to ensure a trouble-free system and reduce problems caused by compressed entrapped air would be to prevent it from entering the system. Methods to eliminate or minimize entrapped air in piping systems are:

1. Slowly fill the piping system (a velocity of 1 fps or less) and vent the air at the piping system's high points.

2. Install continuous acting air relief valves at high points in the line.

3. Install air vents or vacuum relief valves, especially when filling large tanks or other voluminous fluid containers.

ABRASION RESISTANCE/SLURRIES

All piping systems exhibit wear with continued use. However, how a piping material actually wears is a complex phenomenon and depends on many variables, including the following:

Properties of particles—size, percent weight, shape, hardness, specific gravity.

Fluid properties—viscosity, pH, specific gravity.

Pipe-wall properties—hardness, elasticity, composition, density.

Slurry properties—velocities, type of flow.

Changes in any of these variables will increase or decrease the rate of erosion of pipe walls. The ideal slurry solutions are medium velocity, homogeneous flows with larger-sized rounded particles; the most undesirable flow is either saltation or sliding bed flows with fast, angular, small-sized particles traveling at velocities over 10.0 fps. Homogeneous flows are systems in which the solids are uniformly distributed throughout the liquid; saltation flows tend to have solid particles bounce along the pipe bottom; sliding bed flows are the most destructive to piping and comprise solids sliding and rolling along the bottom of the pipe.

Most plastics outperform metal when transporting abrasive slurries. The most successful thermoplastic piping materials are polyethylene (PE 34) and PB. There are many documented cases in which these materials have outlasted steel pipe by as much as 4 to 1. Another material that is designed especially for abrasive services is a ceramic-impregnated lined RTRP. This piping-wall material has proven to be one of the ultimate means of handling severe slurry services. Another thermoplastic material that has excellent abrasive-resistant properties is polyurethane. Small-bore tubing is made of this material, but, to date, no larger solid polyurethane piping material is manufactured. Listed are several abrasive applications successfully handled by plastic piping:

Boiler bottom ash

Mine trailings

Wood chip/pulp

Gravel transport

Fruit/vegetable handling

Dredging

Crushed limestone

Coal slurries

Scrubber solids

Mineral ores

PAINTING PLASTIC PIPING

Any plastic piping system may be painted, taped, or covered for identification, or, on some materials (natural polyolefins, in particular), to prevent ultraviolet light degradation. The key to coating plastic piping systems is not to use any compound that may react with the plastic pipe walls. Piping systems of PVC and CPVC and most reinforced thermoset design systems may have acrylic paint applied directly to the surface with no priming required. Brushing or spraying the paint on is acceptable. For piping materials with high glass and paraffin additives such as PE, PB, PP, and PVDF, a primer is recommended before applying the acrylic paint. One recommended primer that has proven successful with plastic piping applications is Dupont's 329S®. Two other methods of identifying or covering plastic piping are by a thin-walled aluminum or other metal shield and by taping. The tape should be applied using a 50% overlap. Tape-coat® made by Tape-coat of Evanston, IL, has a good reputation for covering plastic piping.

OTHER INSTALLATION DESIGNS

In designing pipelines for submerged and weighted marine use, water surface lines, marsh lines, insert renewal lining, and overland installations, the piping supplier should be contacted and his or her design and installation procedures followed explicitly.

UNDERGROUND PIPING DESIGN

Buried pipelines are subjected to external loads. Plastic pipe, in most instances, is considered a flexible pipe rather than a rigid piping material such as concrete. Rigid pipe is designed to handle internal pressure and all external loads by the strength of the pipe wall itself. Flexible pipe is pipe that is able to bend without breaking and must use the pipe wall and the buried medium to sustain external loads; the pipe and soil form an integral structure. When installed properly, plastic pipe actually gains strength due to the support of the surrounding soil. The soil and pipe wall deflect or compress depending on any one or a combination of three factors:

1. Pipe stiffness,

2. Soil stiffness.

3. Load on the pipe (earth, static, and live loads).

PIPE STIFFNESS

Pipe stiffness is normally defined as the force in psi divided by the vertical deflection in in. Listed in Table 2-7 are typical PVC pipe stiffness factors. The arbitrary datum point of 5% deflection is used as a comparison of pipe stiffness values in different flexible piping. By varying the deflection, the pipe stiffness will vary directly.

SOIL STIFFNESS

Soil stiffness is defined as the soil's ability to resist deflection. Spangler developed the "Modified Iowa Equation" to calculate the deflection of buried flexible pipe in terms of soil stiffness, independent of pipe size. The product of his research resulted in determination of E' values or more commonly called modulus of soil reactions. (Table 2-10 lists the E' values for various soil types.)

EARTH LOADS

Earth loads may be determined using the Marston load formula. Marston's load formula for flexible pipe is

$$W_c = C_d W B_d B_c$$

where

W_c = Load on pipe (lb/linear ft)
W = unit weight of backfill (lb/ft³)
B_c = horizontal width of pipe (ft)
B_d = horizontal width of trench at top of pipe (ft)
C_d = load coefficient
$\quad = \dfrac{1 - e(-2ku'H/B_d)}{2ku'}$
e = natural logarithm base
k = Rankine's ratio of lateral to vertical pressure
u' = coefficient of friction between backfill material and trench walls
H = fill height (ft)

from this equation it can be seen that the wider the trench, the more load will be imposed on the buried pipe. There is a point where increasing the width of the trench does not add any additional load; this point is termed the embankment load.

Table 2-7. PVC Pipe Stiffness (psi)[a]

DR or SDR	Stiffness	
	Minimum E = 400,000 psi	Minimum E = 500,000 psi
42	26	32
41	28	35
35	46	57
33.5	52	65
32.5	57	71
26	115	144
25	129	161
21	234	292
18	364	455
17	437	546
14	815	1,019
13.5	916	1,145

Source: Uni-Bell Plastic Pipe Assn.
[a]Calculations from equation:

$$PS = \frac{F}{DY} = \frac{EI}{0.149\ r^3}$$

$$\text{For PVC pipe:} = 4.47 \frac{E}{(DR - 1)^3}$$

F = force (lb/linear in.)
Y = mean radius of pipe (in.)
t = wall thickness (in.)
E = modulus of elasticity (psi)
I = moment of inertia of the wall cross section per unit length of pipe (in.³)
DR = dimension ratio = OD/t
OD = outside diameter
SDR = standard dimension ratio

The embankment, or sometimes referred to as the prism, load can be determined by multiplying the fill height (H) by the unit weight of backfill by the horizontal width of the pipe:

$$W_c = HWB_c$$

where

W_c = soil load in pipe (lb/linear ft)
H = fill height (ft)
W = unit weight of backfill (lb/ft³)
B_c = horizontal width of pipe (ft)

In Appendix C there are tables shown for soil load calculations using the known trench width calculations ($W_c = C_d W B_d B_c$) and also for the more conservative prism loads ($W_c = HWB_c$). Before determining the static and dynamic soil loads, let us look at trench design.

TRENCH DESIGN

Trenches should be of adequate width to allow the burial of plastic pipe, while being as narrow as practical. If expansion and contraction are not problems and snaking of pipe is not required, minimum trench widths may be obtained by joining the piping outside the trench and then lowering the piping into the trench after testing. A trench width of two or three times the piping diameter is a good rule of thumb in determining the trench width. Listed in Tables 2-8 and 2-9 are suggested minimums for both narrow and supported trench widths.

If the trench width must exceed 6 pipe diameters, the embedment up to the pipe springline must be compacted approximately $2^1/_2$ pipe diameters from each side of the pipe. For pipe diameters above 12 in., the embedment should be compacted 1 pipe diameter or 2 ft on each side of the pipe to the pipe springline. Figure 2-4 shows the pipe springline and other trenching terminology.

The trench bottom should be free of any sharp objects that may cause point loading. Any large rocks, hard pan, or stones larger than $1^1/_2$ in. should be removed to permit a minimum bedding thickness of 4–6 in. under the pipe.

Soil

The type of soil and the amount of compaction of the pipe embedment directly affect the buried pipe's performance. With proper embedment soil and compaction, greater burial depths are possible and higher external pressure capability and less pipe deflection will occur. Table 2-10 lists average modulus (E') of soil types and degree of compaction. Table 2-11 further describes embedment classifications. The maximum height of cover may also be increased when the proper compaction of type of soil is used. Table 2-12 lists the maximum recommended height of cover for plastic piping with a pipe stiffness of 46 lb/in. or more.

Table 2-8. Narrow Trench Width, Minimum

| Nominal Pipe Sizes (diameter in.) | Trench Width, Minimum | |
	Number of Pipe Diameters	Inches
4	4.3	18
6	2.9	18
8	2.9	24
10	2.5	26
12	2.4	30
15	2.0	30
18	1.8	32
21	1.6	34
24	1.5	36
27	1.5	40
30	1.4	42
33	1.4	46
36	1.4	50
40	1.4	56
48	1.3	62

Source: Uni-Bell Plastic Pipe Association.

Table 2-9. Supported Trench Width, Minimum

| Nominal Pipe Sizes (diameter, in.) | Trench Width, Minimum | |
	Number of Pipe Diameters	Inches
4	8.5	36
6	5.7	36
8	4.3	36
10	4.0	42
12	3.4	42
15	3.1	48
18	2.7	48
21	2.4	50
24	2.2	52
27	2.1	56
30	2.0	60
33	1.9	63
36	1.9	68
40	1.8	72
48	1.7	81

Source: Uni-Bell Plastic Pipe Association.

These recommended heights of cover will not exceed $7^1/_2\%$ pipe deflection (if properly installed).

Minimum Cover

There are no fixed rules for minimum burial of plastic piping, but good sense and the following guidelines may be useful:

* Locate pipe below the frostline.

* A minimum cover of 18 in. or one pipe diameter (whichever is greater) when there is no overland traffic.

* A minimum cover of 36 in. or one pipe diameter (whichever is greater) when truck traffic may be expected.

* A minimum cover of 60 in. when heavy truck or locomotive traffic is possible.

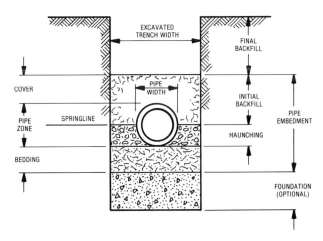

Fig. 2-4. Terminology of trench cross sections. (Courtesy of Uni-Bell Plastic Pipe Association.)

Table 2-10. Average Values of Modulus of Soil Reaction, E'

Soil Type of Initial Backfill Embedment Material	E' for Degree of Compaction of Bedding (psi)			
	Dumped	Slight, < 85% Proctor, <40% Relative Density	Moderate, 85%–90% Proctor, 40%–70% Relative Density	High, >95% Proctor, >70% Relative Density
V	No data available; consult a competent soils engineer; otherwise use E' = 0			
IV	50	200	400	1,000
III	100	400	1,000	2,000
II	200	1,000	2,000	3,000
I	1,000	3,000	3,000	3,000

Adapted from: Uni-Bell Plastic Pipe Association data.

Static Load

One method of calculating soil load has been given. Now let us look at another method of calculating static and dynamic loads. The following methods do not use Marston's load equations to determine load conditions on flexible pipe. The total static load pressure of buried flexible pipe consists of three components:

P_{DE} = the static load pressure of dry or slightly moist soil

P_{WE} = the static load pressure of saturated soil under maximum long-term variable water table

P_B = the static load pressure due to stationary surface structures such as buildings or foundations

The sum of these components equals total static load pressure ($P_s = P_{DE} + P_{WE} + P_B$). To calculate the values of the components, use the following equations:

$$P_{DE} = (H - h)(100 \text{ lb/ft}^3)$$

$$P_{WE} = (h)(130 \text{ lb/ft}^3)$$

$$P_B = \frac{3WZ^3}{2\pi R^5} \quad \text{(Boussinesq Equation)}$$

Table 2-11. Description of Embedment Material Classifications

Soil Class	Soil Type	Description of Material Classification
I		Manufactured angular, granular material, $1/4$–$1\,1/2$ in. (6–40 mm) size, including materials having regional significance such as crushed stone or rock, broken coral, crushed slag, cinders, or crushed shells
II	GW	Well-graded gravels and gravel–sand mixtures, little or no fines; 50% or more of coarse fraction retained on No. 4 sieve; more than 95% retained on No. 200 sieve; clean
	GP	Poorly graded gravels and gravel–sand mixtures, little or no fines; 50% or more of coarse fraction retained on No. 4 sieve; more than 95% retained on No. 200 sieve; clean
	SW	Well-graded sands and gravelly sands, little or no fines; more than 50% of coarse fraction pass No. 4 sieve; more than 95% retained on No. 200 sieve; clean
	SP	Poorly graded sands and gravelly sands, little or no fines; more than 50% of coarse fraction pass No. 4 sieve; more than 95% retained on No. 200 sieve; clean
III	GM	Silty gravels, gravel–sand–silt mixtures; 50% or more of coarse fraction retained on No. 4 sieve; more than 50% retained on No. 200 sieve
	GC	Clayey gravels, gravel–sand–clay mixtures; 50% or more of coarse fraction retained on No. 4 sieve; more than 50% retained on No. 200 sieve
	SM	Silty sands, sand–silt mixtures. More than 50% of coarse fraction pass No. 4 sieve; more than 50% retained on No. 200 sieve
	SC	Clayey sands, sand–clay mixtures. More than 50% of coarse fraction pass No. 4 sieve. More than 50% retained on No. 200 sieve.
IV	ML	Inorganic silts, very fine sands, rock flour, silty or clayey fine sands; liquid limit 50% or less; 50% or more pass No. 200 sieve
	CL	Inorganic clays of low to medium plasticity, gravelly clays, sandy clays, silty clays, lean clays; liquid limit 50% or less; 50% or more pass No. 200 sieve
	MH	Inorganic silts, micaceous or diatomaceous fine sands or silts, elastic silts; liquid limit greater than 50%; 50% or more pass No. 200 sieve
	CH	Inorganic clays of high plasticity, fat clays. Liquid limit greater than 50%; 50% or more pass No. 200 sieve
V	OL	Organic silts and organic silty clays of low plasticity; liquid limit 50% or less; 50% or more pass No. 200 sieve
	OH	Organic clays of medium to high plasticity; liquid limit greater than 50%; 50% or more pass No. 200 sieve
	PT	Peat, muck, and other highly organic soils

Adapted from: Uni-Bell Plastic Pipe Association data.

Table 2-12. Maximum Height of Cover Recommended

Embedment Class	Pipe Zone Condition		Recommended Maximum Height of Cover (ft)
	Percentage of Proctor Density Range	Modulus of Soil Reaction, E' (psi)	
I	–	3000	50
II	85–95	2000	50
	75–85	1000	50
	65–75	200	17
III	85–95	1000	50
	75–85	400	28
	65–75	100	12
IV	85–95	400	28
	75–85	200	17
	65–75	50	9
V	Soil Class Not Recommended		

Source: Uni-Bell Plastic Pipe Association.

where

H = depth of soil from ground level to top of pipe (ft) (see Fig. 2-5)

h = depth of saturated wet soil above pipe (ft) (see Fig. 2-5)

100 lb/ft³ = approximate weight of dry soil

130 lb/ft³ = approximate weight of water-saturated soil

W = superimposed surface load (lb) (see Fig. 2-6)

z = vertical distance from the point of load (A) to the top of pipe (ft)

R = straight-line distance from point of load (A) to the top of pipe (ft)

$$R = \sqrt{x^2 + y^2 + z^2}$$

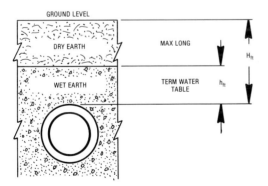

Fig. 2-5. Schematic diagram of terms used in static load calculations. (Courtesy of Phillips Drisco Pipe.)

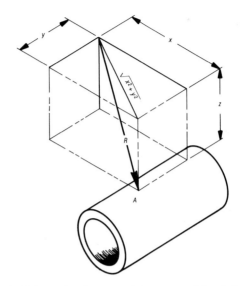

Fig. 2-6. Schematic diagram of terms used in static load calculations. (Courtesy of Phillips Drisco Pipe.)

Table 2-13. Pipe Wall Dimensions of Schedule 40 and 80 Plastic Pipe

Nominal Pipe Size[a] (in.)	Outside Diameter 40 & 80 Pipe (in.)	Schedule 80		Schedule 40	
		Avg. Inside Diameter (in.)	Minimum Wall (in.)	Avg. Inside Diameter (in.)	Minimum Wall (in.)
1/8	0.405	0.203	0.095	0.261	0.068
1/4	0.540	0.288	0.119	0.354	0.088
3/8	0.675	0.407	0.126	0.483	0.091
1/2	0.840	0.528	0.147	0.608	0.109
3/4	1.050	0.724	0.154	0.810	0.113
1	1.315	0.935	0.179	1.033	0.133
1 1/4	1.660	1.256	0.191	1.364	0.140
1 1/2	1.900	1.476	0.200	1.592	0.145
2	2.375	1.913	0.218	2.049	0.154
2 1/2	2.875	2.289	0.276	2.445	0.203
3	3.500	2.864	0.300	3.042	0.216
4	4.000	3.786	0.337	3.998	0.237
5	5.563	4.767	0.375	5.017	0.258
6	6.625	5.709	0.432	6.031	0.280
8	8.625	7.565	0.500	7.943	0.322
10	10.750	9.492	0.593	9.976	0.365
12	12.750	11.294	0.687	11.890	0.406

[a]Larger pipe sizes are available.

Table 2-14. Dimensions of PVC C-900 and SDR[a] 35 Sewer Pipe

Nominal Pipe Size (in.)	PVC C-900 Pipe				PVC Sewer Pipe	
	Outside Diameter (in.)	SDR25 Minimum Wall (in.)	SDR18 Minimum Wall (in.)	SDR14 Minimum Wall (in.)	Outside Diameter (in.)	SDR35 Minimum Wall (in.)
4	4.800	0.192	0.267	0.343	4.215	0.120
6	6.900	0.276	0.383	0.493	6.275	0.180
8	9.050	0.362	0.503	0.646	8.400	0.240
10	11.100	0.444	0.617	0.793	10.500	0.300
12	13.200	0.528	0.733	0.943	12.500	0.360
15	NCA[b]	NCA	NCA	NCA	15.300	0.430

[a]SDR = standard dimension ratio.
[b]NCA = not commonly available.

Table 2-15. Minimum Wall Thicknesses (in.) of

Nominal Pipe Size (in.)	ABS				CPVC		Polybutylene				
	145 psi	180 psi	230 psi	Airline 185 psi	Fire Sprinkler SDR 13.5	CTS	SDR 45.5 45 psi	SDR 26 80 psi	SDR 21 100 psi	SDR 17 125 psi	SDR 13.5 160 psi
$1/2$	NCA	NCA	0.091	0.090	NCA	0.068	NCA	NCA	NCA	NCA	0.062
$3/4$	NCA	NCA	0.105	0.090	0.083	0.080	NCA	NCA	NCA	NCA	0.072
1	0.086	0.105	0.132	0.114	0.103	0.102	NCA	NCA	NCA	NCA	0.091
$1 1/4$	NCA	0.132	0.165	NCA	0.130	0.125	NCA	NCA	NCA	NCA	0.120
$1 1/2$	0.124	0.151	0.189	0.177	0.149	0.148	NCA	NCA	NCA	NCA	0.140
2	0.155	0.189	0.236	0.222	0.186	0.193	NCA	NCA	NCA	NCA	0.176
$2 1/2$	NCA	NCA	NCA	0.266	0.226	NCA	NCA	NCA	NCA	NCA	0.259
3	0.227	0.278	0.347	0.317	0.274	NCA	NCA	NCA	NCA	NCA	0.333
4	0.292	0.357	0.447	0.386	NCA	NCA	0.099	0.173	0.214	0.265	0.491
6	0.430	0.526	NCA	NCA	NCA	NCA	0.146	0.255	0.315	0.390	0.639
8	0.559	NCA	NCA	NCA	NCA	NCA	0.190	0.332	0.411	0.507	0.796
10	NCA	NCA	NCA	NCA	NCA	NCA	0.236	0.413	0.512	0.632	0.944
12	NCA	NCA	NCA	NCA	NCA	NCA	0.280	0.490	0.607	0.750	1.037

Table 2-15 (Concluded). Minimum Wall Thicknesses (in.) of Commonly Used Plastic Pipe (Other than Schedule 40 or 80)[a]

Nominal Pipe Size (in.)	PVC					PVDF	RTRP			
	SDR 21 200 psi	SDR 26 160 psi	SDR 32.5 125 psi	SDR 41 100 psi	SDR 64 63 psi	150–230 psi	Filament Wound Lined	Filament Wound Unlined	Centrifugal Cast Lined	Centrifugal Cast Unlined
$1/2$	NCA	NCA	0.045	NCA	NCA	0.070	0.231	NCA	NCA	NCA
$3/4$	0.062	NCA	0.045	NCA	NCA	0.070	0.208	NCA	NCA	NCA
1	0.063	0.060	0.050	NCA	NCA	0.090	0.072	NCA	NCA	NCA
$1 1/4$	0.079	0.064	0.060	NCA	NCA	0.090	NCA	NCA	NCA	NCA
$1 1/2$	0.090	0.073	0.060	NCA	NCA	0.110	0.082	NCA	0.250	0.175
2	0.113	0.091	0.073	NCA	NCA	0.120	0.115	0.070	0.250	0.200
$2 1/2$	0.137	0.110	0.088	NCA	NCA	0.100	NCA	NCA	0.250	0.200
3	0.167	0.135	0.108	0.086	NCA	0.110	0.115	0.075	0.250	0.200
4	0.214	0.173	0.138	0.110	0.070	0.140	0.115	0.095	0.250	0.200
6	0.316	0.255	0.204	0.162	0.104	0.200	0.145	0.123	0.300	0.250
8	0.410	0.332	0.265	0.210	0.135	0.240	0.158	0.143	0.300	0.250
10	0.511	0.413	0.331	0.262	0.168	0.310	0.188	0.170	0.300	0.250
12	0.606	0.490	0.392	0.311	0.199	0.390	0.218	0.200	0.300	0.250

[a]Notes:

1. Listed values are averages and may vary from one manufacturer to another, especially for RTRP.

2. Larger diameter pipe is available in PVC, polyolefins and RTRP materials.

3. Abbreviations: CTS = copper tube size (this pipe has the same OD as copper pipe); IPS = iron pipe size; NCA = not commonly available; SDR = standard dimension ratio.

x and y = horizontal distances at 90° to each other from point of load (A) to the top of pipe (ft)

Live Load

Live or dynamic loads are caused by moving vehicles or heavy equipment or both. The shallower the flexible pipe is buried, the more concern there is about live loads. Usually pipe below 10 ft of cover will not be affected significantly by dynamic loads. If the application prevents deep burial of the pipe (pipe buried less than 10 ft) and there will be heavy traffic passing over the pipe, it would be prudent to use a steel or reinforced-concrete casing to prevent damage to the plastic pipeline. To calculate live load pressure, use the Boussinesq equation, but multiply the superimposed dynamic load (W) by $1 1/2$.

Total External Soil Pressure

The total external soil pressure is the total of static and live pressure loads plus the addition of any vacuum (in lb/ft^2) in the piping system. A vacuum or negative pressure causes compressive hoop stress in the pipe wall and assists the forces trying to collapse the pipe.

Commonly Used Plastic Pipe (Other than Schedule 40 or 80)ᵃ

SDR 11 200 psi CTS/IPS	SDR 9 250 psi	Polyethylene (PE 3408)						Polypropylene		PVC SDR 17 250 psi
		SDR 32.5 50 psi	SDR 26 65 psi	SDR 21 80 psi	SDR 17 100 psi	SDR 13.5 130 psi	SDR 11 160 psi	150 psi	45 psi	
NCA	NCA	NCA	NCA	NCA	0.060	0.060	NCA	0.100	NCA	NCA
0.080 CTS/NCA	0.097	NCA	NCA	0.060	0.060	0.072	0.080	0.100	NCA	0.062
0.102 CTS/NCA	0.125	NCA	NCA	0.060	0.070	0.091	0.102	0.120	NCA	0.077
0.125 CTS/NCA	0.153	NCA	NCA	0.073	0.092	0.120	0.125	0.140	NCA	0.098
0.148 CTS/NCA	0.181	NCA	NCA	0.085	0.107	0.140	0.148	0.180	NCA	0.112
0.193 CTS/0.216	0.264	NCA	NCA	0.109	0.138	0.180	0.193	0.230	NCA	0.140
0.318 IPS	0.389	NCA	NCA	NCA	NCA	NCA	NCA	0.270	NCA	0.169
0.409 IPS	0.500	0.108	0.135	0.167	0.206	0.259	0.318	0.320	NCA	0.206
0.602 IPS	0.736	0.138	0.173	0.214	0.265	0.333	0.409	0.390	0.140	0.265
0.784 IPS	0.958	0.204	0.255	0.315	0.390	0.491	0.602	0.580	0.200	0.390
0.977 IPS	NCA	0.265	0.332	0.411	0.507	0.639	0.784	0.720	0.240	0.508
1.159 IPS	NCA	0.331	0.413	0.512	0.632	0.796	0.977	0.900	0.310	0.632
1.273 IPS	NCA	0.392	0.490	0.607	0.750	0.944	1.159	1.130	0.390	0.750

Summary

To summarize, the primary factors that influence the performance of flexible pipe are pipe stiffness, soil stiffness, and static/live loads. The vertical force of earth and live load causes pipe to deflect and transmit load to the soil support on the side of the pipe. The stiffness of the pipe and soil resist deflection and will achieve, over time, an equilibrium of forces acting on the pipe and soil.

DEFLECTION

Deflection is essential for designing a suitable long-lasting flexible buried piping system. Of course, there is a limit of deflection desired. In most plastic piping systems, a 5% deflection is the maximum designed application. This does not mean the pipe will fail if this value is exceeded, but rather there might be a small change in the cross-sectional area of the pipe that will affect flow velocities. To calculate the deflection of plastic pipe, Spangler's modified Iowa equation is used:

$$\Delta X = D_L \frac{KW_c r^3}{EI + 0.061 E' r^3}$$

where

ΔX = horizontal deflection, in.
D_L = deflection lag factor (use 1.25 as normal)
K = bedding constant (use 0.1 as normal)
W_c = load per unit length of pipe, lb/ linear in.

r = mean pipe radius, in.
E = modulus of tensile elasticity of the pipe material, psi
I = moment of inertia per unit length, in.³
E' = modulus of soil reaction, psi

The relationship between horizontal deflection (ΔX) and vertical deflection (ΔY) in buried flexible pipe was determined by Spangler in the following equation:

$$\Delta X = 0.913 \Delta Y$$

where

ΔX = horizontal deflection, in.
ΔY = vertical deflection, in.

WALL BUCKLING

Wall buckling is seldom the limiting factor in designing for buried flexible pipe, but it should be calculated, especially for nonpressurized pipelines. P_{CB} is the critical-collapse pressure and P_t is the total soil pressure, and to prevent buckling P_{CB} must be larger than P_t:

$$P_{CB} = 0.8\sqrt{E' P_c}$$

where

E' = soil modulus
P_c = critical-collapse differential
$= \dfrac{2.32(E)}{(SDR)^3}$
E = modulus of elasticity of pipe
SDR = standard dimension ratio of pipe

Table 2-16. Weights of Commonly

Nominal Pipe Size (in.)	ABS					CPVC						
	Sch 40 DWV	145 psi	180 psi	230 psi	Airline 185 psi	Sch 80	Sch 40	Fire Sprinkler SDR 13.5	SDR 11 CTS	SDR 45.5 45 psi	SDR 26 80 psi	SDR 21 100 psi
1/2	NCA	NCA	NCA	0.090	0.850	0.228	0.180	NCA	0.086	NCA	NCA	NCA
3/4	NCA	NCA	NCA	0.150	0.130	0.308	0.239	0.168	0.140	NCA	NCA	NCA
1	NCA	0.150	0.190	0.230	0.168	0.453	0.352	0.262	0.240	NCA	NCA	NCA
1 1/4	0.310	NCA	0.300	0.370	NCA	0.624	0.475	0.418	0.360	NCA	NCA	NCA
1 1/2	0.370	0.330	0.390	0.480	0.272	0.760	0.568	0.548	0.460	NCA	NCA	NCA
2	0.500	0.510	0.610	0.750	0.365	1.050	0.761	0.859	0.860	NCA	NCA	NCA
2 1/2	NCA	NCA	NCA	NCA	0.579	1.602	1.201	1.257	NCA	NCA	NCA	NCA
3	1.030	1.100	1.330	1.620	0.758	2.146	1.572	1.867	NCA	NCA	NCA	NCA
4	1.470	1.820	2.190	2.680	1.079	3.126	2.239	NCA	NCA	0.590	1.000	1.200
6	2.625	3.940	4.740	NCA	NCA	6.077	3.945	NCA	NCA	1.300	2.200	2.700
8	NCA	6.670	NCA	NCA	NCA	9.263	NCA	NCA	NCA	2.200	3.700	4.500
10	NCA	NCA	NCA	NCA	NCA	13.886	NCA	NCA	NCA	3.300	5.800	7.100
12	NCA	NCA	NCA	NCA	NCA	NCA	NCA	NCA	NCA	4.700	8.100	9.900

Table 2.16 (*Concluded*). Weights of

Nominal Pipe Size (in.)	Polypropylene		PVC							
	150 psi	45 psi	Sch 80	Sch 40 DWV	SDR 17 250 psi	SDR 21 200 psi	SDR 26 160 psi	SDR 32.5 125 psi	SDR 41 100 psi	SDR 64 63 psi
1/2	0.080	NCA	0.208	0.162	NCA	NCA	NCA	0.074	NCA	NCA
3/4	0.120	NCA	0.280	0.219	0.122	0.120	NCA	0.093	NCA	NCA
1	0.170	NCA	0.411	0.320	0.162	0.160	0.155	0.130	NCA	NCA
1 1/4	0.260	NCA	0.569	0.431	0.260	0.250	0.210	0.198	NCA	NCA
1 1/2	0.400	NCA	0.690	0.519	0.330	0.354	0.296	0.227	NCA	NCA
2	0.630	0.230	0.957	0.693	0.521	0.543	0.451	0.346	NCA	NCA
2 1/2	0.980	0.380	1.460	1.133	0.761	0.798	0.649	0.506	NCA	NCA
3	1.280	0.470	1.950	1.455	1.131	1.155	0.956	0.755	0.625	NCA
4	1.910	0.710	2.844	2.050	1.870	1.913	1.561	1.241	0.995	0.660
8	4.040	1.490	5.433	3.615	4.052	4.167	3.394	2.700	2.158	1.398
6	6.300	2.280	8.251	5.436	6.510	7.069	5.778	4.567	3.643	2.360
10	9.900	3.610	12.243	7.713	10.400	11.035	9.001	7.109	5.664	3.665
12	15.690	5.710	16.831	10.195	14.949	15.599	12.728	9.986	7.974	5.148

[a]Values are averages and may vary from one manufacturer to another, especially with RTR pipe. In most PVC, polyolefins and RTR piping materials, larger diameter
SDR = standard dimension ratio; DR = dimension ratio; RTRP = reinforced thermosetting resin pipe; DWV = drain, waste, and vent applications.

SUMMARY

In general, when proper trench design, backfill, and compaction of 85% or better density is applied, wall crushing, buckling, and deflection are rarely encountered. The design engineer, in non-pressurized applications, should calculate the safety factors and ensure short- and long-term performance of the buried pipe when the pipe is subjected to compressive external forces. When the pipe is pressurized, the internal pressure normally exceeds any external pressures.

PLASTIC PIPING WEIGHTS AND DIMENSIONS

Plastic pipe is lighter in weight than other materials used in piping. Because of its light weight, plastic is easy to handle. Some piping materials are lighter than water and, for marine applications, can be easily floated into place before underwater burial.

The wall thickness for plastic piping varies with

Used Plastic Pipe (lb/ft)[a]

Polybutylene				Polyethylene (PE3408)						Polypropylene	
SDR 17 125 psi	SDR 13.5 160 psi	SDR 11 200 psi	SDR 9 250 psi	SDR 32.5 50 psi	SDR 26 65 psi	SDR 21 80 psi	SDR 17 100 psi	SDR 13.5 130 psi	SDR 11 160 psi	Sch 80	Sch 40
NCA	0.050	NCA	NCA	NCA	NCA	NCA	0.056	0.051	NCA	0.140	0.110
NCA	0.094	NCA	0.099	NCA	NCA	0.074	0.073	0.089	0.152	0.189	0.140
NCA	0.139	NCA	0.165	NCA	NCA	0.108	0.109	0.143	0.243	0.271	0.210
NCA	0.248	NCA	0.245	NCA	NCA	0.184	0.187	0.248	0.418	0.379	0.277
NCA	0.335	NCA	0.345	NCA	NCA	0.251	0.254	0.337	0.589	0.448	0.340
NCA	0.550	0.630	0.590	NCA	NCA	0.411	0.418	0.555	0.935	0.623	0.460
NCA	NCA	NCA	NCA	NCA	NCA	NCA	NCA	NCA	NCA	NCA	NCA
NCA	1.100	1.400	NCA	0.490	0.610	0.740	0.910	1.120	1.350	1.266	0.910
1.500	1.900	2.300	NCA	0.810	1.000	1.230	1.500	1.850	2.230	1.852	1.320
3.300	4.100	4.900	NCA	1.750	2.170	2.660	3.250	4.030	4.840	3.599	2.300
5.600	6.900	8.300	NCA	2.970	3.680	4.520	5.500	6.820	8.210	5.500	3.580
8.600	10.100	12.900	NCA	4.620	5.710	7.010	8.550	10.590	12.750	8.200	5.070
12.100	15.000	18.100	NCA	6.480	8.040	9.860	12.030	14.890	17.940	10.950	6.710

Commonly Used Plastic Pipe (lb/ft)[a]

C 900 DR 25 Class 100	C 900 DR 18 Class 150	C 900 DR 14 Class 200	Sewer/ Drain SDR 35	PVDF		RTRP			
				Sch 80	150 psi	Filament Wound Lined	Filament Wound Unlined	Centrifugal Cast Lined	Centrifugal Cast Unlined
NCA	NCA	NCA	NCA	0.244	0.140	0.200	NCA	NCA	NCA
NCA	NCA	NCA	NCA	0.330	0.180	0.400	NCA	NCA	NCA
NCA	NCA	NCA	NCA	0.487	0.300	0.600	NCA	NCA	NCA
NCA	NCA	NCA	NCA	NCA	0.380	NCA	NCA	NCA	NCA
NCA	NCA	NCA	NCA	0.814	0.550	0.950	NCA	0.850	0.630
NCA	NCA	NCA	NCA	1.126	0.730	1.150	0.400	1.150	0.850
NCA	NCA	NCA	NCA	NCA	0.740	1.450	NCA	1.320	1.230
NCA	NCA	NCA	NCA	2.300	1.000	1.900	0.700	2.170	1.870
1.920	2.610	3.300	1.100	3.365	1.520	2.150	1.100	2.400	2.390
3.960	5.410	6.840	2.400	6.540	2.440	3.600	2.300	4.230	4.220
6.810	9.320	11.760	4.200	NCA	4.910	5.100	3.300	6.140	5.840
10.250	14.000	17.710	6.600	NCA	7.800	6.850	4.800	7.360	7.350
14.500	19.800	25.040	9.400	NCA	12.350	10.100	6.700	8.660	8.370

pipe is available. Abbreviations: NCA = not commonly available; CTS = copper tube size (this pipe has the same OD as copper tube); Sch = schedule;

diameter and schedule or pressure rating of the pipe. Owing to current manufacturing techniques, it is possible for wall thicknesses to vary, but, in thermoplastic piping, manufacturers must meet minimum wall thicknesses as specified by ASTM and other standard-setting organizations. Thermoset piping varies more widely in weights and wall dimensions, owing to the type of construction and diverse manufacturing process of each major producer.

Table 2-13 lists the outside and inside diameters and wall thicknesses for schedule 40 and 80 pipe. Table 2-14 lists the outside diameters and wall thicknesses of AWWA C-900 and sewer pipe. Table 2-15 lists the minimum wall thicknesses of nonschedule pipe. Table 2-16 lists the average weights for most commonly used plastic piping materials. All of these tables are averages, and dimensions and weights may vary slightly in the case of thermoplastic piping. Reinforced thermosetting pipe diameters and weights vary much more widely. To determine more exacting data, request the pipe weights and dimensions from the manufacturer.

3
Joining Techniques

One of the major advantages of plastic pipe is the variety of proven joining systems that may be used. The same joining technique will not work with all materials, but for most plastic piping materials there is a minimum of three types of joints to choose from.

Particular manufacturers have their own special joining instructions, which may differ slightly from those presented in this chapter. If in doubt, follow the manufacturer's instructions.

All joints require three initial steps:

1. Inspection of material to be joined.

2. Cutting the pipe.

3. Cleaning of pipe and fittings.

INSPECTION

It is essential to spend a few minutes before installing plastic piping carefully inspecting the pipe and fittings for any flaws, including deep scratches, excessive warping, broken pipe ends, fine cracks on or under the surface of the plastic material (commonly called "crazing"), delamination (the separation of layers) of RTRP, or any other obvious physical defect. Thermoplastic pipe working pressure can be severely affected if the pipe is deeply gouged or scratched. If there is any concern as to possible damage, play safe and cut out a potentially defective piece of pipe or discard a defective fitting. Crazing is a possible defect of RTRP. A sharp blow by an object to the exterior of the pipe can cause the glass wrappings to withdraw from the resin-rich inner layer of the pipe

and weaken the pipe. This phenomenon is sometimes called "star crazing," owing to the impact point's resemblance to a star (see Fig. 3-1). To observe this defect, check the exterior and, if possible, the interior of the pipe. If there is a whitish spot emanating tentacles of white, this could be crazing, and the crazed section must be cut out of the pipe.

CUTTING

Plastic pipe should be cut with plastic cutting tools. There are excellent tools for thermoplastics that eliminate the need for saws, files, or sandpaper. Every thermoplastic pipe should be cut square, deburred, and then beveled as shown in Figs. 3-2 and 3-3. This procedure has a threefold purpose:

1. The square cut allows the pipe to bottom-out to the socket of the fitting or bell.

2. If cementing or fusion-sealing pipe, any burrs or sharp edges of pipe may scrape off cement or molten plastic when being pushed into the socket.

3. Beveling the pipe allows for easier installation into socket depths and acts as a repository for cement or melted plastic.

For smaller diameter thermoplastic pipe (6 in. or less), a special cutting tool is available. This cutter (Fig. 3-2A) is shown mounted on a 2-in. section of Schedule 80, PVC pipe. During cutting, it is rotated evenly through two to eight turns, de-

Fig. 3-1. The white arrow points out the typical appearance of "star crazing" on a section of fiberglass reinforced plastic (RTR) pipe. (Courtesy of PPS-Maryland, Inc.)

pending on the material, thickness, and size of the thermoplastic pipe. A handy plastic pipe deburrer, shown in Fig. 3-2C, is also available, constructed with consecutive beveling lips for a range of smaller pipe sizes. This deburring and beveling tool is fitted against the freshly cut square edge of the pipe (Fig. 3-2B) and is twisted three or four times. The result is the 2-in. pipe shown at the left in Fig. 3-2C, made with a clean cutting, deburring, and beveling operation that took less than a minute.

For larger diameter thermoplastic pipe (6–28 in.) large cutters are available for in trench or aboveground cutting. Outboard rollers keep the cutter aligned on the pipe to ensure a square cut while simultaneously producing a 15° bevel on the plastic pipe. Carbide-tipped blades penetrate up to 2-in. pipe wall thickness (Fig. 3-3A).

Another tool specifically designed for large diameter piping is a joiner for gasketed and solvent

cemented piping. The tool with saddles for IPS and water pipe outside diameters can join pipe from 4 to 16 in. (Fig. 3-3B).

If plastic pipe cutters are not available, a power or hand hacksaw, or a circular or band saw may be used. For best results, use fine-tooth blades (16–18 teeth per inch) and little or no set (maximum 0.025 in.). A speed of about 6,000 ft/min is suitable for circular saws. A speed of 3,000 ft/min is suitable for band saws. Carbide-tipped saw blades are preferable when quantities of pipe are to be cut. To ensure square-end cuts, a mitre box, holddown, or jig should be used.

For cutting RTR pipe, the following tools have been used successfully:

1. Hand hacksaw (22–28 teeth/in.).

2. Power saw with aluminum oxide abrasive wheel.

3. A band saw (16–22 teeth/in. at speeds of 200–600 ft/min).

CLEANING

Most thermoplastic joints are of a "homogeneous monolithic" nature. This means that the joint has the same physical and chemical characteristics as the pipe and fitting, with no stratification. The joining surfaces should be free of dirt, grease, water, mold release, or other foreign substances. If the surfaces cannot be successfully cleaned with a clean cloth, use emery cloth or sandpaper. Some manufacturers, in fact, prefer that the joining areas be sanded and cleaned before joining. The manufacturer's instructions should be read thoroughly to determine the proper joint preparation.

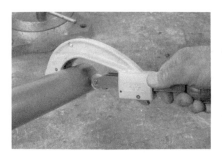

A

B

C

Fig. 3-2. (A) Use a specially designed pipe cutter for thermoplastic pipe of less than 6 in. diameter. Turn pipe cutter evenly, two to eight turns, depending on the plastic material and the diameter of the pipe. (B) Apply a bevel to the straight cut pipe with three or four turns of a deburrer. (C) A plastic pipe deburrer will remove any burrs and bevel the square-cut pipe (shown at the left already beveled). Cutting and beveling a 2-in. PVC Schedule 80 pipe should take less than 1 min.

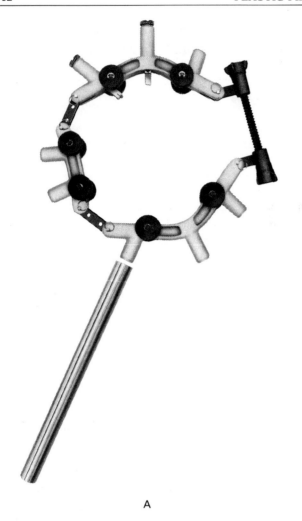

A

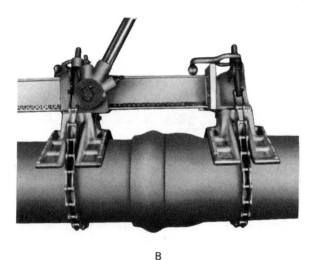

B

Fig. 3-3. (A) Cutter for large-diameter thermoplastic pipe. (B) Pipe joiner for large-diameter pipe. (Courtesy of Reed Manufacturing.)

JOINING PREPARATION—POINTS TO REMEMBER

1. Before joining, always carefully inspect pipe fittings and valves for any external damage.

2. All plastic pipe should be cut as square as possible.

3. Wherever possible, use cutting tools designed for plastic pipe.

4. Deburr and/or bevel all plastic pipe before joining.

5. Clean the surface ends of all pipe and fittings to be joined.

6. Do not clean plastic pipe with solvents—use a clean cloth or sandpaper.

JOINING INSTRUCTIONS

This section will give detailed instructions on how to join pipe using various joining methods. All the methods work if done correctly, and they have been proven in many field installations. Photographs have been used extensively to demonstrate the proper techniques. Any deviations from the stated instructions should be carefully considered and reviewed with the product manufacturer or supplier. Table 3-1 outlines the various joining methods available for plastic materials; it also lists the advantages and disadvantages of each.

SOLVENT CEMENTING

By far, solvent cementing is the most popular method of joining aboveground thermoplastic piping systems. The outside diameter of the pipe and the inside diameter of the fitting are coated with cement and joined together. In a matter of seconds after the pipe and fitting are joined, the bond is formed, and the pipe and fitting cannot be separated.

Not all thermoplastics can be solvent cemented. The advantages of this joining method are that (1) no special tools are needed and (2) the joint produced is strong. The major disadvantage is that seconds after the pipe and fitting are joined, the joint cannot be changed or moved nor can the system be fully tested until the newly cemented joint is completely cured.

Plastic Systems Applicable—For PVC, CPVC,

Table 3-1. Joining Methods for Plastic Pipe and Fittings

Joining Method	Material Applicable	Advantages	Disadvantages
Solvent—cemented	PVC, CPVC, ABS, styrene	Quick set-up; strongest possible joint; no foreign substance contacting fluid being handled; no special tools required; the most used and experience-proven plastic joint.	Cannot be taken apart once joint is made; cannot join under poor weather conditions; leaks cannot be fixed easily; system cannot be tested immediately.
Adhesive bonding (bell and spigot)	RTRP	Very strong joint; the best proven method of joining small diameter RTR pipe; no alignment problems; use of tapered joints allows movement of pipe before full curing.	Limited "pot life" of adhesive; must have good weather conditions to join; difficult to fix joint links; system cannot be tested immediately unless heat is applied.
Adhesive bonding butt-strap	RTRP	Very strong joint; excellent flow characteristics; low thermal expansion; lightweight; much faster joining time than welded pipe.	Must have excellent environment conditions; on large-diameter pipe, inside and outside of pipe must be done; difficult to align pipe; cannot test system immediately unless auxiliary heat source is used; time-consuming when compared to other plastic joints.
Butt heat fusion	Polyethylene, polypropylene, polybutylene, PVDF	(Same as Socket heat fusion)	(Same as Socket heat fusion)
Electrical resistance heat fusion	Polypropylene	Can "dry fit" entire system; relatively quick installation time; leaks can be fixed relatively easily; can fuse two joints at a time with one machine	Need an electrical source; requires expensive tool (machine); possibility of contamination of fluid by coil; coil damage is hard to detect.
Socket heat fusion	Polypropylene, polyethylene, PVDF polybutylene	No foreign materials contacting fluid being handled; very strong joint; once the tools are prepared, a quick joining method; quick setup; can be tested immediately.	Difficult to use in cold weather conditions; special tooling required; need electrical or gas source; cannot be taken apart once joint is made.
Flanging	All	(Same as Threaded)	High initial cost; dissimilar material contacting fluid being handled; limited to 150 psig working pressure; not compact dimensionally compared to other joints.
Threaded	All	Easy to disassemble; can join dissimilar materials; can prefabricate a system for field installation; if a leak exists, can possibly fix easily; system can be tested immediately.	Tendency to leak above 2 in. IPS; reduces working pressure of system; threads are weakest part of piping system; difficult to thread above 2 in. IPS; requires special tools for threading.
O-ring	PVC, polyethlene, polybutylene, RTRP	Easy to install in all weather conditions; no special tools required; can test system immediately; excellent pressure ratings; allows some misalignment in piping system; self-compensating for expansion and contraction in pipe.	Dissimilar material contacting fluid being handled; not recommended for fluids other than water; possible "hang up" of sediment with O-ring; thrust blocking required; not recommended above ground.
Compression insert	Polyethylene, polybutylene	Easy to install; no special tools required; easy to assemble and disassemble; easy to fix leaks.	Fittings are different material than pipe; limited to maximum of 3-in.-diameter pipe; poor flow characteristics
Grooved	PVC, CPVC, ABS, PVDF	Easy to assemble and disassemble; can prefabricate a system for field installation; can test the system immediately; allows some misalignment in piping systems.	Expensive tools required; foreign material in contact with liquid being handled; couplings relatively expensive; may reduce optimum pressure rating of pipe.
Mechanical joint (grooving and compression)	Polypropylene (DWV)	Easy to assemble and disassemble; easy to fix leaks; no expensive tools required.	O-ring or "olive" is different material than pipe and fittings, limited to maximum of 4-in.-diameter pipe.

styrene, and ABS, in pressure and drainage systems.

Instructions—For PVC and CPVC—other plastics are similar, but might have slightly different manufacturer's recommendations.

1. Using a *natural bristle* brush of the correct width (see Table 3-2) apply a complete coating of primer to the entire outside surface of the pipe end and to the mating inside surface of the connection socket. (See Fig. 3-4).

Table 3-2. Suggested Brush Size for Cementing Pipe

Nominal Pipe Size (in.)	Ideal Brush Size[a]	
	Maximum Width (in.)	Minimum Length (in.)
$1/4$	$1/2$	1
$3/8$	$1/2$	1
$1/2$	$1/2$	1
$3/4$	$1/2$	$1^1/2$
1	1	$1^1/2$
$1^1/4$	1	$1^1/2$
$1^1/2$	$1^1/2$	$1^3/4$
2	$1^1/2$	2
$2^1/2$	2	$2^1/2$
3	$2^1/2$	3
4	3	$3^1/2$
5	4	$4^1/2$
6	5	$5^1/2$
8	6	6

[a]It is recognized that the ideal brush size may not always be readily available. The selection should come as close as possible to the ideal in order to ensure complete coverage with a minimum number of brush strokes. Furthermore, the containers for the primer and solvent cement should be wide enough to accommodate the width of the brushes.

2. Before the primer dries, use another natural bristle brush of the same width to brush solvent cement *liberally* as follows:

a. *On the pipe*—brush *liberally* once around the entire surface of pipe OD to a width slightly more than the socket depth of the fitting. (See 3-5A.)

b. *On the fitting*—brush *lightly* but completely around the entire depth of the socket surface. (See Fig. 3-5B.)

c. *On the pipe*—apply another liberal coating of cement, as before.

3. *Immediately* upon finishing the cement application, insert the pipe into the *full* socket depth while rotating the pipe or fitting a quarter turn to ensure complete and even distribution of the cement. (See 3-6A.)

4. Hold joint together for a minimum of 10–15 sec to ensure that the pipe does not back out of the socket.

5. *Immediately* after joining, wipe all excess cement from the surface of the pipe and fittings including any globs of cement that may have been dropped onto the pipe or fitting. (See Fig. 3-6B.)

Solvent Cementing—Points to Remember

1. Do not attempt to cement wet surfaces. (Do not cement in the rain.)

2. Pipe, fittings, and solvent cement should

Fig. 3-4. Applying primer to end of pipe.

A

B

Fig. 3-5. (A) Applying solvent cement to the pipe. (B) Applying solvent cement to the connection socket.

be exposed to the same temperature for at least 1 hr prior to cementing.

3. Use only *natural* bristle brushes. Primer and cement will dissolve synthetic bristles.

A

B

Fig. 3-6. (A) Inserting pipe into connection socket. (B) Cleansing away excess cement.

4. Do not cement in weather below 40° F (4.4° C).

5. Above 90° F (32° C), and under direct exposure to the sun, cement as follows:

a. Shade joint surfaces from the sun's rays for a minimum of 1 hr prior to joining and continue joining in the shade.
b. Use a clean cotton rag or dish swab instead of a brush to apply the primer.
c. Use a plunger-type safety can to store the primer.
d. Apply cement quickly and insert pipe into socket as quickly as possible after applying cement.

6. Schedule 40 cement may be used for Schedule 40 and SDR pipe and fittings, 2 in. and under.

7. Schedule 80 cement must be used for Schedule 80 (inclusive), Schedule 40, and SDR pipe and fittings over 2 in.

8. Do not use cements designed for a particular plastic with another plastic material. (It may or may not work.) There are general-purpose cements, but check with the pipe and fitting manufacturer's recommendations, particularly on pressure applications.

9. CPVC cement should be used for all CPVC joints, with no exceptions.

10. Cement only plastic materials that can be solvent-welded. For example, polypropylene, PVDF, polyethylene, and polybutylene cannot be properly joined using cements.

11. On 6-in.-diameter pipe and above, especially in hot weather, two workers should apply cement to the pipe while one worker is applying cement to the fitting in order to minimize application time, thereby avoiding premature setting in earlier coats of cement.

12. Do not discard empty primer or cement cans near plastic pipe.

13. Observe the "use prior to" date on cement.

14. If cement becomes lumpy and stringy, throw it away. Do not attempt to thin out sluggish cement with thinner or primer. *Throwing away potentially ineffective cement is less costly than fixing a leak.*

15. Appropriate joint-drying time should elapse before the cemented joint is moved or subjected to internal or external pressure. Table 3-3 shows suggested drying times before movement.

16. Listed in Table 3-4 is a guide for the total amount of joints that may be made from a par-

Table 3-3. PVC and CPVC Joint Movement Times

Nominal Pipe Sizes (in.)	Hot Weather[a] [90–150° F (32–66° C) Surface Temperature]	Mild Weather[a] [50–90° F (10–32° C) Surface Temperature]	Cold Weather[a] [10–50° F (−12–32° C) Surface Temperature]
¹/₂–1¹/₄	12 min	20 min	30 min
1¹/₂–2¹/₂	30 min	45 min	1 hr
3–4	45 min	1 hr	1¹/₂ hr
6–8	1 hr	1¹/₂ hr	2¹/₂ hr

[a]Some manufacturers recommended 48 hr for joint-drying time before moving the cemented joint.

Table 3-4. Number of Joints per Volume of Cement and Primer

Pipe Size (in.)	1 Pint Primer 1 Pint Cement	1 Quart Primer 1 Quart Cement	1 Gallon Primer 1 Gallon Cement
$1/2$	190	380	1520
$3/4$	120	240	960
1	100	200	800
$1^1/4$	70	140	560
$1^1/2$	50	100	400
2	30	60	240
$2^1/2$	25	50	200
3	20	40	160
4	16	32	128
5	12	24	96
6	6	12	48
8	2	4	16

ticular volume of cement and primer, relative to pipe diameter.

17. When cementing PVC or CPVC ball valves, carefully cement the pipe in the valve with the ball closed. This will prevent ball damage and will ensure operation of the valve.

18. *Do not take short cuts—follow instructions completely.*

BELL AND SPIGOT ADHESIVE BONDING

This joint is used for RTR piping where the pipe is joined into a fitting socket after an adhesive is applied to the pipe's outside diameter and to the inside diameter of the fitting. The adhesive bonds the pipe and fitting together through a chemical parafin. The resultant joint is stronger than either the pipe or the fitting. This is the preferred method of joining RTR piping systems, if belled or socket fittings are available, since it minimizes alignment problems that may exist in the butt and strapping method of RTRP joining.

Plastic Systems Applicable—For all reinforced thermoset resin pipe (RTRP).

Instructions—Manufacturer's bonding procedures differ in that some have tapered-end pipe and, others, straight-end pipe. The following procedure assumes that the pipe has been tapered, if required:

1. Remove all gloss from pipe end approximately $1/2$ in. farther than the depth of the fitting socket. As shown in Fig. 3-7A, use of emery cloth is one method of abrasion.

2. Remove sanding dust with a clean cloth being careful not to contaminate the sanded surface (Fig. 3-7B).

3. Mix hardener and adhesive thoroughly to manufacturer's proportions. Remember that "pot life," the time span between the mixing of the adhesive and when it is no longer usable, is only 15 or 20 min to 1 hr, depending on temperature

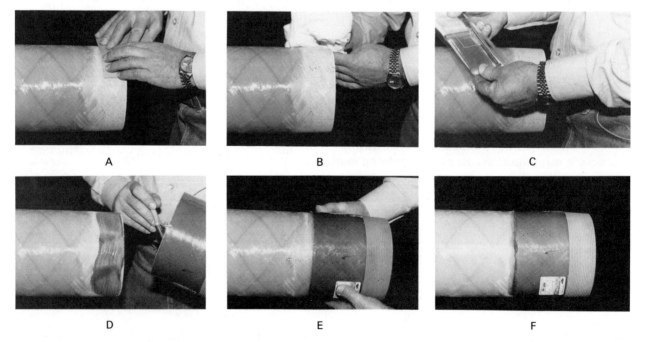

A   B   C

D   E   F

Fig. 3-7. Steps in bell and spigot adhesive-bonding procedure. (Courtesy of PPS-Maryland, Inc.)

and humidity conditions. (A two-part mixing bag, supplied by the manufacturer, is shown in Fig. 3-7C.)

4. Depending on the consistency of the adhesive cement mix, it may be applied with a smooth stick, a putty knife, or a natural bristle brush.

5. Coat adhesive to the socket surface of the fitting approximately $1/16$ in. thick, but to the outer surface of the pipe apply the adhesive more generously (about $1/8$–$3/16$ in. thick) (Fig. 3-7D).

6. Firmly push fitting onto end of pipe, using a stabbing motion then turning one-half turn, as far as it will go (Fig. 3-7F).

7. Any excessive adhesive forced out around the pipe at the socket entrance should be formed into a fillet (Fig. 3-7F).

Bell and Spigot Adhesive Bonding—Points to Remember

1. Check pot life of adhesive with manufacturer's recommendations.

2. In cold weather [70° F (21° C), or below] prewarm the adhesive kits and bell and spigot.

3. Avoid contact with the adhesive and hardener, since they are capable of causing skin and eye irritation.

4. Use of a heat gun or electric collar can shorten cure time of joints.

5. Follow the adhesive manufacturer's mixing instructions explicitly.

6. Maintain back axial pressure (compression) on all previously assembled joints to reduce the chance of separation.

7. After joining, tapered pipe and fittings may be moved without misalignment or separation more quickly than nontapered pipe.

8. In hot weather, when greater separate quantities of adhesive and hardener are supplied than are necessary for immediate use, pot life can be maintained by mixing properly proportioned amounts of hardener and adhesive in paper cups, and using these smaller made-ready quantities. A plastic cooler chest can also be used in hot weather to extend pot life.

BUTT-STRAP ADHESIVE BONDING

This method of joining fiberglass-reinforced pipe consists of butting pipe end-to-end and then wrapping various layers of resin-impregnated glass over the butted pipe ends. The joint is exceedingly strong, but careful alignment and a good environment must exist to utilize this joining technique.

Plastic Systems Applicable—For all reinforced thermoset resin pipe (RTRP).

Instructions

1. Using a disk sander (note the use of a safety mask), rough the outside surfaces approximately 1 in. farther than the finished weld. When inside bonds are used, inside surfaces should also be sanded prior to assembly. (See Fig. 3-8A.)

2. With a clean cloth wipe away the sand from ends of pipe to be joined.

3. Using a natural bristle brush, coat raw end edge of pipe or duct with a small amount of catalyzed resin. Silica-filled resin paste should be used to fill large voids.

4. Pipe ends should be butted together tightly, and all pipe should be firmly supported to prevent movement during the bond-cure period. This is important, particularly with larger sections. To simplify the initial fixing of pipe ends, the "hot patch" technique is utilized, during which the butting pipe sections are supported as rigidly as possible. With hot patching, small squares of mat are wetted with a small amount of triple-catalyzed resin. (A small amount of resin using three times the normal catalyst should be prepared in a separate container, then discarded immediately after use to avoid contaminating the resin for the weld. Do not attempt to use this material in the preparation of the final adhesive bond.) The hot patches, so prepared, will cure or harden in a matter of minutes and are used quickly and pressed out with a roller to remove air bubbles and to "tack" the surfaces into proper alignment.

5. Once the hot patches are in place, saturate the C-veil strip of glass composition (furnished in the bonding kit) with resin, then center it and wrap it carefully around the joint to be bonded. The C-veil is applied, then brushed over with added resin only at this point in the procedure; it acts as a base on which to construct the bond.

Fig. 3-8. Steps in butt-strap adhesive bonding. (Courtesy of Gulf Wandes.)

Each additional weld layer is composed of three reinforcement strips of varying widths, the widest strip at the bottom and progressively narrower strips laid on top of each other. Each strip is also positioned so the bottom, wider one, will overlap at both sides of the preceding strip by 1 in.; thus, a "feather edge" layer results when the joint is made. (To estimate the widths of the overlays, see Table 3-5.) On a clean piece of kraft

wrapping paper lay out the three required types of glass or fabric mat strips in the proper sequence. Normally, the strips of glass or fabric mat are supplied already cut to size in varying widths and are approximately 2 in. longer than the pipe circumference. (On circumferences larger than 16 in., the weld strips may be cut in half-lengths to simplify application.) Finally, mix only the amount of resin that can be used in 20–30 min.

6. To produce the first bond layer, with a paint roller or brush spread a thin layer of resin over the widest strip on the kraft paper. (See Fig. 3-8B.)

7. Center the next widest strip on top of the first, starting 1 in. in from each side to provide the stepped construction shown in Fig. 3-8C.

Table 3-5. Government Voluntary Standard Pipe Wall Thicknesses and Required Widths of Adhesive Bonding Overlays

Pipe Wall Thickness (in.)	$3/16$	$1/4$	$5/16$	$3/8$	$7/16$	$1/2$	$9/16$	$5/8$	$11/16$	$3/4$
Minimum total width of overlay (in.)	3	4	5	6	7	8	9	10	11	12

Wet out this second strip with resin in the same way as the first. When this is done, follow the same procedure with the third, and narrowest, strip (Fig. 3-8D). Be sure that all three strips are thoroughly saturated with resin.

8. Once you have laid out the entire first layer of the bond, compress it with the corrugated aluminum roller to remove large air bubbles. This will also help to work the resin thoroughly into the reinforcement strips.

9. Next, starting at one end, peel the entire bond layer from the kraft paper. NOTE: Be sure that all three reinforcement strips are stuck together and come off as *one* layer, as shown in Fig. 3-8E.

10. Carefully center this first layer of the bond on the joint—WITH THE NARROWEST STRIP ON THE INSIDE AGAINST THE PIPE.

11. Wrap the bond layer around the joint until the free end overlaps the beginning. Be sure that the bond is centered and tight. Care should be taken to avoid wrinkles on the under or back side of bonds. Once this first bond layer is in place, brush on additional resin and roll it thoroughly with the small, corrugated aluminum roller to work out trapped air and rough spots (Fig. 3-8F). Add additional bond layers—one layer at a time—using the procedure previously described, until the required total number of reinforcement strips (recommended by the manufacturer) has been applied.

12. With the complete bond thus in place, position a section of surface veil (the thick, paperlike material) over the bond, apply resin, and work out air bubbles and rough spots with the roller. Working from the center of the bond, roll toward the edges to produce a smooth, flawless structure. (See Figs. 3-8G and 3-8H.) The surface veil is used to avoid direct contact with the wet resin and assists in producing a smooth joint.

13. If preferred, one-mil-thick plastic films, such as Saran wrap, may be used in place of the surface veil. However, a slight change in technique will be necessary. Wrap the completed bond with the film, allowing approximately $1/2$ in. overlap on both sides. Work out the air with a diagonal motion using a rubber spatula, which has first been dipped in uncatalyzed resin or other lubricant such as glycerine.

14. The bond (weld) should be left physically undisturbed until hard, to avoid misalignment. Welding should not be done when ambient temperatures are below 65° F (18° C). Temporary heat, such as infrared lamps, may be used to offset lower temperatures. For exterior installation, it may be necessary to protect the weld from inclement weather.

Butt-Strap Adhesive Bonding—Points to Remember

1. Mix only as much resin as is required for immediate use.

2. Store catalyst away from all sources of heat.

3. Always add catalyst into the resin and stir (never the reverse).

4. Excess catalyst and used-paper working surfaces should be immediately discarded or burned in a safe area.

5. Proper alignment of pipe ends is critical to ensure a good joint.

6. After making joints, clean skin with warm water and a commercial detergent.

7. The particular manufacturer will usually recommend the total number of strips required per bond. There are more layers required the higher the pressure rating, and larger strip widths with greater pipe diameters.

BUTT HEAT FUSION

Butt heat fusion consists of heating and melting pipe ends of the same material and pressing the melted surfaces together. The subsequent cooling bonds the two pipes together with a joint equal to or exceeding the strength of the pipe itself. This joining method is used principally in large-diameter (over 4 in.) polyolefin piping.

The principal advantages of this joining method are the excellent joint strength and the ease in joining; the disadvantage is the costly joining equipment.

Plastic Systems Applicable—For polypropylene, polyethylene, polybutylene, and PVDF.

Instructions

1. **Turn butt fusion unit on**—Assemble all required tools and set temperature of heater plate to manufacturer's recommendations (Fig. 3-9A). When not in use, the heater plate should always be returned to this holder.

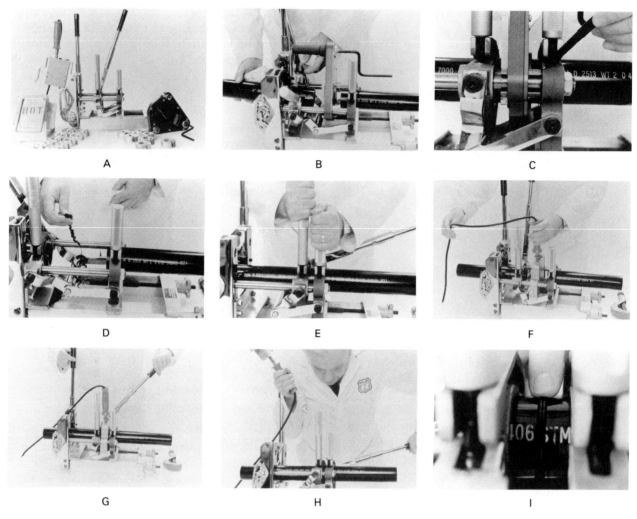

Fig. 3-9. Steps in joining pipe by the butt heat fusion method. (Courtesy of Philipps Driscopipe.)

2. **Secure pipe ends**—Pipe ends and/or fittings are placed and secured in the joining device clamping shells (Fig. 3-9B). One shell is movable; the shorter length of pipe is clamped in this shell. The longer length of pipe, one with previous joints in it, is secured in the other shell, the stationary one. The pipe sections to be joined should protrude out of the shells 2–6 in., depending on the size of pipe and estimated amount of out-of-squareness of the ends.

3. **Position facer**—A rotating facer is positioned between the secured pipe ends. This facer is free to slide axially (Fig. 3-9C).

4. **Face pipe ends**—By moving the movable shell and its pipe end against the facer (Fig. 3-9D) and the sliding facer against the end of the pipe protruding out of the stationary shell and maintaining a moderate pressure against the rotating facer, both pipe ends are almost simultaneously faced until they are square.

5. **Check squareness**—When the facing operation is completed, the facer feels or sounds to be running free and the pigtails, or shavings, are long and continuous in length, the pipe ends are then backed off and the facer is removed. The pipe ends should be moved together and butted against one another in order to visually check their relative squareness and smoothness. UNDER NO CIRCUMSTANCES SHOULD THE FACED ENDS OF THE PIPE BE TOUCHED; touching faced ends may cause contamination of the joint. (See Fig. 3-9E.)

6. **Match outside pipe end diameters**—Since differences in pipe ovality or tolerances be-

tween two pieces of pipe become very pronounced in pipe sizes over 12 in., it is necessary to use the adjustable matching clamps, an integral part of the shells, to adjust the two outside diameters of the pipe ends until they match smoothly.

7. **Note machine drag (optional)**—Pipe producers who recommend following definite melt cycle and joining pressures also recommend that a machine drag (pressure it takes to move the movable shell with the pipe length clamped in it) be determined and added to their recommended melt cycle and joining pressures. If there is a pressure gage or a torque wrench on the butt fusion unit, this machine drag reading is noted or read after the pipe ends have been faced and as the movable shell is moved forward to check the squareness of the faced ends. (See Fig. 3-9F.)

8. **Perform melt cycle**—After the faced ends are checked for squareness, the pipe ends are backed off from one another; the heater plate, which like the facer can slide axially, is inserted between them. The faced pipe ends are then butted up against the heater plate in the same manner that they were butted up against the facer. The pipe producer's recommendation is followed as to the amount of pressure that the pipe ends exert against the heater and the length of time that this pressure should be maintained, or manufacturer's recommended pressure sufficient for continuing melt to form a uniform $1/16$–$1/8$ in. melt bead on the outside edge of the pipe ends. (See Figs. 3-9G and 3-9H.)

9. **Heat soak the pipe ends**—When the designated time (as recommended by most manufacturers) has elapsed and/or the desired size and type of melt bead has developed, all pressure by the pipe ends against the heater face is completely relieved, but contact is maintained for the purpose of allowing the heat to soak into the pipe and walls without forming a larger melt bead. The soaking time may vary from 5 sec to several minutes, depending on the type of plastic, pipe size, and wall thickness.

10. **Join the melted pipe ends**—When the designated soaking time has elapsed, the pipe ends are *quickly* snapped away from the heater, the heater is *quickly* removed and the melted pipe ends are *quickly* and firmly slammed together,

to ensure proper mixing of the melts and forming the weld bead as shown in Fig. 3-9I. Follow producer's recommendation as to the pressure and length of time or to the physical characteristics of the weld bead.

11. **Cooling Times**—The pipe may be unclamped and carefully removed from the machine after it has dropped to a temperature of 150–160° F (66–71° C), and the hand can be held comfortably on the joint. This will vary in time from 30 sec to 45 min. If joining pressure was held during this entire cooling time, then the pipe can be immediately removed as soon as the pressure has been relieved. After the pipe has returned to ambient temperature, the joint can be subjected to whatever stresses the wall of the pipe will withstand.

Butt Heat Fusion—Points to Remember

1. Keep the heater faces clean.

2. Wear gloves and avoid contact with the heater and any hot melted material.

3. Cut the pipe ends as square as possible in order to reduce facing time or to eliminate the need for refacing with prior readjustment of the pipe end in the shell.

4. Try to perform the joining operations on a string of pipes by moving the pipe rather than the machine.

5. During the winter, especially on windy days, shelter the heater and the joining process.

6. When in doubt about the quality of a joint, cut it out and redo it. This is especially applicable for pipe that is to be inserted or buried, or is to be used in underwater installation where location and repair of a leak will be very costly.

ELECTRICAL RESISTANCE FUSION

This joining method is a socket heat fusion method in which the heat necessary to melt the outside diameter of the pipe and the inside diameter of the fitting is developed by electrolyzed copper coils. The coiled wire remains an integral part of the completed socket-fused joint.

Fuseal and Lab-Line®,* polypropylene acid waste drainage product lines, are at present the

---

*Fuseal is a trademark of the R. & G. Sloane Mfg. Co.; Lab-Line is a trademark of the Enfield Industrial Corp.

only joining methods of this type in the country. The advantages and disadvantages of the method are similar to other heat fusion joints.

Plastic Systems Applicable—For polypropylene acid-waste drainage and pressure systems.

Instructions

1. Place the fusion coil on the OD of the pipe, pigtail-end first, leaving the opposite end of the coil flush with the end of the pipe (Fig. 3-10A). (Lab-Line's® coil is fixed in the fitting; this procedure bypasses steps 1 and 2.)

2. Both pipe and coil are inserted into the fitting socket. The pipe penetrates all the way into the socket while the coil stops against the counterbore shoulder (Fig. 3-10B).

3. A compression clamp is tightened over the end of the socket. The terminal block from the power unit is then slipped into the yoke provided on the clamp and the pigtails are connected to the binding posts of the terminal block (Figs. 3-10C and 3-10D).

4. The voltmeter is set to the correct size pipe, in this case $1\frac{1}{2}$ min, as recommended by the manufacturer (6 in. size requires 3 min).

5. About 30 sec before the timer runs out, the clamp is retightened firmly to create extra bonding pressure.

6. When the proper time has elapsed, attachments are removed, and care is taken not to damage or remove the pigtails (Fig. 3-10F).

Electrical Resistance Fusion—Points to Remember

1. Do not snip off the pigtails until the joints have been tested. If a leak occurs, additional current may seal the joint.

2. Do not tighten compression clamp excessively, since this could damage the fitting or pipe or both.

3. When adjusting the power unit, do not have a coil connected to the terminal block.

4. When making many joints, dry-fit the entire system, secure all the compression clamps, and, by using several power units, make the joints simultaneously.

5. When using the power unit in temperatures below 65° F (18° C), it might be necessary to increase the voltage time over $1\frac{1}{2}$ min.

SOCKET HEAT FUSION

Socket heat fusion is the heating, melting, and pressing together of the outside diameter of the pipe end and the inside diameter of a fitting socket.

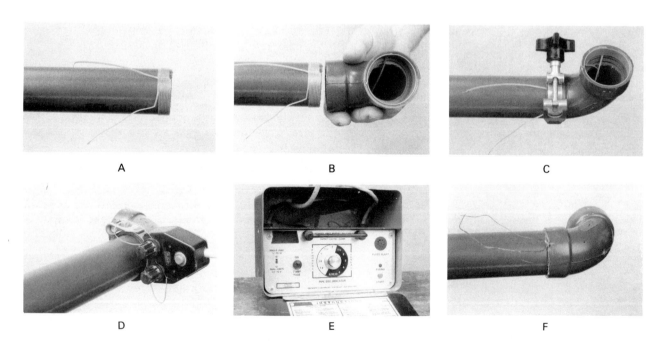

A   B   C

D   E   F

Fig. 3-10. Making joints using the electric resistance heat fusion method.

After cooling, the finished joint is stronger than either the pipe or the fitting. Small-diameter polypropylene, polyethylene, polybutylene, and PDVF pressure piping systems use this method of joining. The advantages and disadvantages are similar to that of the butt heat fusion method of joining.

Plastic Systems Applicable—For polypropylene, polybutylene, and PVDF pressure and polyethylene acid-waste drainage.

Instructions

1. Install the male and female tool pieces on either side of the sealing tool, which is used for this joining method, and secure with set screws. (See Fig. 3-11A.)

2. Insert the electrical plug into a grounded 110 Vac electrical source and allow the tool to come to the proper operating temperature. The tool temperature is read directly from the mounted temperature gage, and the tool temperature can be adjusted by turning the thermostat adjustment screw with a screwdriver. (See Fig. 3-11B.)

3. Place the proper size depth gage over the end of the pipe. (See Fig. 3-11C.)

4. Attach the depth gaging clamp to the pipe by butting the clamp up to the end of the depth and locking it into place. Then remove the depth gage (See Fig. 3-11D.)

5. Simultaneously place the pipe and fitting squarely and fully on the heat tool pieces so that the ID of the fitting and the OD of the pipe are in contact with the heating surfaces. Care should be taken to ensure that the pipe and fitting are not cocked when they are inserted on the tool pieces. (See Fig. 3-11E.)

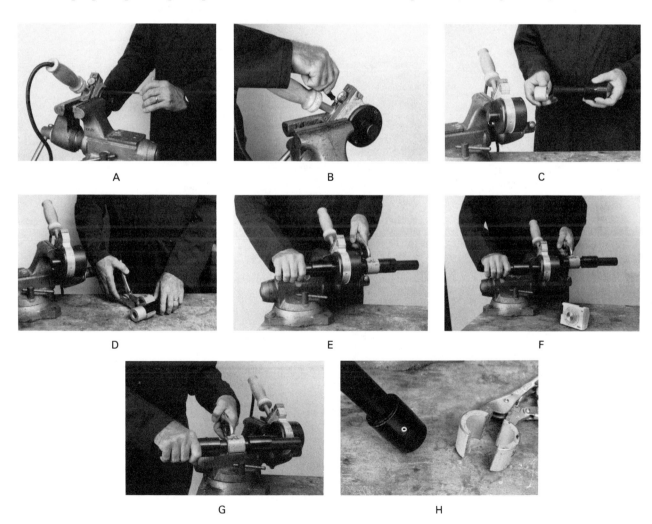

Fig. 3-11. Making joints using the socket heat fusion method. (Courtesy of Chemtrol Div. of Nibco.)

6. Hold the pipe and fitting on the tool pieces for the prescribed amount of time. During this time a bead of melted material will appear around the complete circumference of the pipe at the entrance of the tool piece. (See Fig. 3-11F.)

7. Simultaneously remove the pipe and fitting from the tool pieces and immediately insert the pipe, squarely and fully, into the socket of the fitting. Hold the completed joint in place and avoid relative movement between components for at least 15 sec. Once a joint has been completed, the clamp can be removed and preparation for the next joint can be started. (See Fig. 3-11G.)

8. The surfaces of the male and female tool pieces are Teflon coated to prevent sticking of the hot plastic. (See Fig. 3-11H.)

Socket Heat Fusion—Points to Remember

1. Keep the tool pieces clean by removing residue after making each joint.

2. Respray contact surfaces with silicone spray after each joint is made.

3. Use recommended heating temperatures for particular pipe materials as suggested by the manufacturer. Wear gloves and avoid contact with hot plastic material or heating element of tool.

4. If making joints indoors, have good ventilation to remove gases that may be emitted from the plastic material.

5. If making joints outdoors, try to protect the installation area from wind and cold temperatures so that the tool will heat up quickly and evenly.

6. Do not test until the joint has fully cooled to ambient temperature.

7. When putting pipe in the fitting, do not turn pipe.

FLANGING

There is no plastic piping system available that cannot be flanged. Normally, a flange joint is used for the following reasons:

1. For a temporary piping system.

2. Must connect to another noncompatible piping system.

3. To minimize field labor and field expertise in joining.

4. Where environmental conditions negate other joining methods.

The principal disadvantages of the flange joining method are the rather high initial material costs and special handling required.

Plastic Systems Applicable—For all plastic piping systems.

Instructions

1. Use soft, full-face gaskets (approximately $1/8$ in. thick) compatible to the liquid going through the pipe (Fig. 3-12A).

2. Install the flange, making certain the pipe is fully "bottomed-out" to the flange socket depth (Fig. 3-12B).

3. Bolts should be tightened by pulling down on the nuts diametrically opposite each other with a torque wrench. (See Figs. 3-12C and 3-12D.) Recommended bolt torques are shown in Table 3-6.

4. All bolts should be pulled down gradually to a uniform tightness. For additional flange data, see Table 3-7.

Flanging—Points to Remember

1. Teflon full-face gaskets are not recommended. (Teflon envelope gaskets may be used.)

2. Make certain the pressure rating of the flange is not exceeded.

3. Do not overtighten bolts, as cracking of the plastic flange can occur.

4. Washers should be used on both flanges to be joined.

5. Use well-lubricated bolts.

6. Carefully follow recommended bolt torques.

7. Do not use ring gaskets; use full-face gaskets only.

THREADING

Threaded plastic systems are normally not recommended in (1) systems above 2 in., owing to

Fig. 3-12. Steps in installing plastic flanges.

possible out-of-roundness of threads; (2) high-pressure environments; and (3) where leaks can be dangerous to personnel. The advantages of a threaded system are that it is temporary and that it can be used to join plastics to nonplastic materials.

Plastic Systems Applicable—For all thermoplastic materials Schedule 80, or heavier, and 4 in. IPS (Iron Pipe Size), and smaller.

Table 3-6. Recommended Torques For Well-Lubricated Bolts

Flange Size (in.)	Bolt Diameter (in.)	Torque (ft/lb)
$^1/_2$–$1^1/_2$	$^1/_2$	10–15
2–4	$^5/_8$	20–30
6–8	$^3/_4$	33–50
10	$^7/_8$	53–75
12	1	80–110

Table 3-7. Plastic Flange Data

Iron Pipe Size	Flange Diameter (in.)	Number of Holes	Diameter of Holes (in.)	Bolt Diameter (in.)[a]	Flange Thickness (in.)[b]
$^1/_2$	$3^1/_2$	4	$^5/_8$	$^1/_2$	$^7/_{16}$
$^3/_4$	$3^7/_8$	4	$^5/_8$	$^1/_2$	$^1/_2$
1	$4^1/_4$	4	$^5/_8$	$^1/_2$	$^9/_{16}$
$1^1/_4$	$4^5/_8$	4	$^5/_8$	$^1/_2$	$^5/_8$
$1^1/_2$	5	4	$^5/_8$	$^1/_2$	$^{11}/_{16}$
2	6	4	$^3/_4$	$^5/_8$	$^3/_4$
$2^1/_2$	7	4	$^3/_4$	$^5/_8$	$^{15}/_{16}$
3	$7^1/_2$	4	$^3/_4$	$^5/_8$	1
4	9	8	$^3/_4$	$^5/_8$	$1^1/_8$
6	11	8	$^7/_8$	$^3/_4$	$1^1/_4$
8	$13^1/_2$	8	$^7/_8$	$^3/_4$	$1^3/_8$
10	16	12	1	$^7/_8$	$1^1/_2$
12	19	12	1	$^7/_8$	2

[a]Length of bolt = 2 × flange thickness + bolt diameter + $^3/_8$ in. (use U.S. Standard round washers).
[b]Flange thickness dimensions are for molded thermoplastic, 150-psi flanges. RTR flanges differ dimensionally in thickness.

Instructions

1. Use threading dies and vises designed for plastic pipe. Be sure the vise is tight enough, yet, does not damage the plastic pipe.

2. Insert a wood or aluminum plug well into the pipe end. This prevents distortion of pipe walls to avoid cutting off-center threads (Fig. 3-13A).

3. Turn threading dies slowly, keeping speed constant. A standard thread-cutting oil, silicone water, or soap-and-water solution could be used as a lubricant (Fig. 3-13B).

4. Before threading plastic to plastic, or plastic to metal, use Teflon tape or Teflon lubricant. Teflon acts as a lubricant and sealer. Tape should completely cover all threads; overlap each wrap $1/4$ in. Rectorseal® in. is a sealer that has been successfully used in many field applications (Fig. 3-13C).

5. Screw fittings onto pipe and tighten with strap wrenches. Avoid excessive torque; one to two threads past hand-tight is adequate (Fig. 3-13D).

Threading—Points to Remember

1. Only Schedule 80 or heavier-walled plastic pipe should be threaded.

2. Plastic threading equipment is recommended; however, brass threading equipment may be used. Dies must be clean and sharp with a 5–10° negative front rake angle.

3. Do not use pipe wrenches or pliers on plastic pipe or fittings. Teeth marks from metal tools

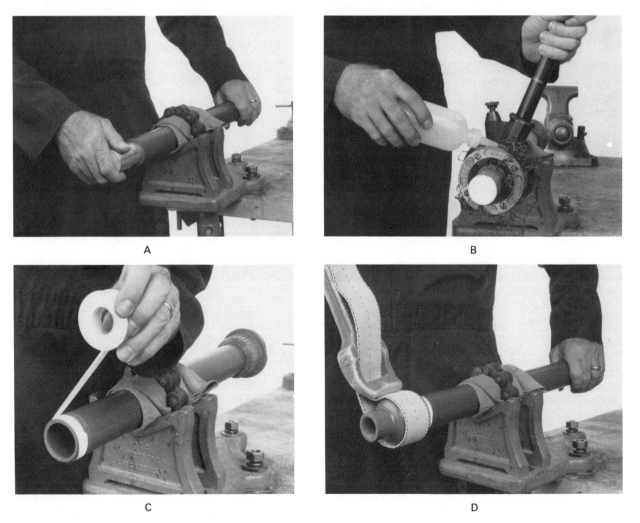

A

B

C

D

Fig. 3-13. Threading plastic pipe.

are a sure sign that incorrect tools were used to tighten plastic threaded joints.

4. Do not subject threaded joints to repeated or severe angular strain or vibrations.

5. Threaded plastic pipe reduces the pipe operating pressure approximately 50%.

6. When threading plastics, use a sealer and lubricant at all times.

7. Do not, if at all possible, use plastic threaded joints above four inches IPS (Iron Pipe Size).

8. Tables 3-8 and 3-9 give useful drill and tap information.

O-RINGS

An O-ring joint, mostly used in underground systems handling water, is made by inserting a circular, ringlike elastomer into a grooved fitting. It prevents leakage and minimizes "backing-out" of the pipe from the grooved fitting. Underground pressure-rated piping is the dominant product line that uses this type of joint. The advantage of this joint is the ease in joining. The disadvantage is that it tends to leak more readily than other joints and, consequently, is not recommended for aboveground pressure applications or transporting contaminating fluids below ground.

Plastic Systems Applicable—For underground pressure-rated pipe.

Instructions

1. Clean the ring groove in pipe or fitting. Make certain no dirt or foreign material interferes with the proper seating of the ring.

Table 3-8. Drill Sizes for Pipe Taps

Size of Tap (in.)	Number of Threads Per Inch	Diameter of Drill	Size of Tap (in.)	Number of Threads Per Inch	Diameter of Drill
$^1/_8$	27	$^{11}/_{32}$	2	$11^1/_2$	$2^3/_{16}$
$^1/_4$	18	$^7/_{16}$	$2^1/_2$	8	$2^5/_8$
$^3/_8$	18	$^{37}/_{64}$	3	8	$3^1/_4$
$^1/_2$	14	$^{23}/_{32}$	$3^1/_2$	8	$3^3/_4$
$^3/_4$	14	$^{59}/_{64}$	4	8	$4^1/_4$
1	$11^1/_2$	$1\,^5/_{32}$	$4^1/_2$	8	$4^3/_4$
$1^1/_4$	$11^1/_2$	$1\,^1/_2$	5	8	$5^5/_{16}$
$1^1/_2$	$11^1/_2$	$1^{23}/_{32}$	6	8	$6^3/_8$

Table 3-9. Tap and Drill Sizes (American Standard Coarse)

Size of Drill	Size of Tap	Threads Per Inch	Size of Drill	Size of Tap	Threads Per Inch
7	$^1/_4$	20	$^{49}/_{64}$	$^7/_8$	9
F	$^5/_{16}$	18	$^{53}/_{64}$	$^{15}/_{16}$	9
$^5/_{16}$	$^3/_8$	16	$^7/_8$	1	8
U	$^7/_{16}$	14	$^{63}/_{64}$	$1\,^1/_8$	7
$^{27}/_{64}$	$^1/_2$	13	$1\,^7/_{64}$	$1\,^1/_4$	7
$^{31}/_{64}$	$^9/_{16}$	12	$1\,^7/_{32}$	$1\,^3/_8$	6
$^{17}/_{32}$	$^5/_8$	11	$1^{11}/_{32}$	$1\,^1/_2$	6
$^{19}/_{32}$	$^{11}/_{16}$	11	$1^{29}/_{64}$	$1\,^5/_8$	$5^1/_2$
$^{21}/_{32}$	$^3/_4$	10	$1\,^9/_{16}$	$1\,^3/_4$	5
$^{23}/_{32}$	$^{13}/_{16}$	10	$1^{11}/_{16}$	$1\,^7/_8$	5
			$1^{25}/_{32}$	2	$4^1/_2$

2. Set ring in groove with manufacturer's markings showing the proper orientation. Smooth the ring so that it sets evenly all around and is free from twists (Fig. 3-14A).

3. Clean the entire circumference of the pipe from the end to 1 in. beyond the reference mark. (See item 6.) Make certain the pipe end is properly beveled; if it is not, use a plastic beveling/deburring tool to prepare the pipe end (Fig. 3-14B).

4. Lubricate the spigot end of the pipe using hand, cloth, pad, sponge, or glove. The lubricant used should be to manufacturer's recommendations. The lubricant coating should be the equivalent thickness of a brush coat of paint.

5. Insert the bevel pipe end into the bell so that the end is in contact with the ring. The pipe lengths being joined are held close to the ground to keep the lengths in proper alignment (Fig. 3-14C).

6. Push the spigot end in until the reference mark on the spigot end is flush with the end of the bell. The proper reference mark (see Table 3-10) should be indicated with a pencil or crayon.

O-Rings—Points to Remember

1. Do not lubricate the rubber ring or groove in the bell, since the lubricant could cause ring displacement.

Table 3-10. Pipe End Referencing for O-Ring Joining

Pipe size (in.)	$1^1/_2$	2	$2^1/_2$	3	4	6	8
E dimension (in.)	3	$3^1/_4$	$3^1/_2$	$3^3/_4$	$4^1/_4$	$5^1/_4$	5.9

Source: Johns-Manville Corp.

A   B   C

Fig. 3-14. Proper method of installing O-ring joints. (Courtesy of Uni-bell Plastic Pipe Assoc.)

2. Use proper lubricant, since nonapproved lubricant may harbor bacteria or damage the rings.

3. If undue resistance to insertion of the bevel end is encountered, disassemble the joint and check the ring position. If it is twisted or pushed out of its seat, clean the ring, bell, and bevel end and repeat the assembly steps.

4. Install ring in the proper orientation or it will not seal.

COMPRESSION INSERT

Small-diameter polyethylene and polybutylene (where the pipe wall is slightly flexible) are the only materials that can use the compression-insert type of joining technique. A compression-insert joint consists of a barbed fitting inserted into the inside diameter of the pipe. Normally, a clamping device is attached to the outside diameter of the pipe and pressure is applied.

Plastic Systems Applicable—For polyethylene and polybutylene flexible coiled pipe.
Instructions

1. Use insert fittings of PVC, styrene, nylon, or metal, depending on application, with stainless steel or cadmium-coated clamps (Fig. 3-15A).

2. Slip the clamp (or clamps) on first, then insert the fitting to shoulder depth (Fig. 3-15B).

3. Position the clamp(s) directly over the serrations and tighten with a screwdriver. On $1\frac{1}{4}$ in. and larger-size pipe, double clamping is recommended.

4. When a connection to a threaded fitting on an existing line is made, screw the insert adapter into the existing connection before slipping the pipe over the adapter (Fig. 3-15C).

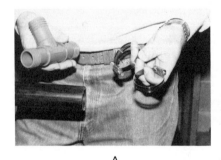

A   B   C

Fig. 3-15. Installing compression insert fittings.

Compression Insert—Points to Remember

1. To aid insertion of the fitting, especially in cold weather, dip the end of the pipe in hot water, which will soften it. Do not use any lubricant other than water.

2. When screwing plastic insert fittings, use a strap wrench to tighten the fitting.

3. Kinked pipe should not be used.

4. Do not use ordinary pipe joint compounds when joining threaded plastic to metal, or plas-

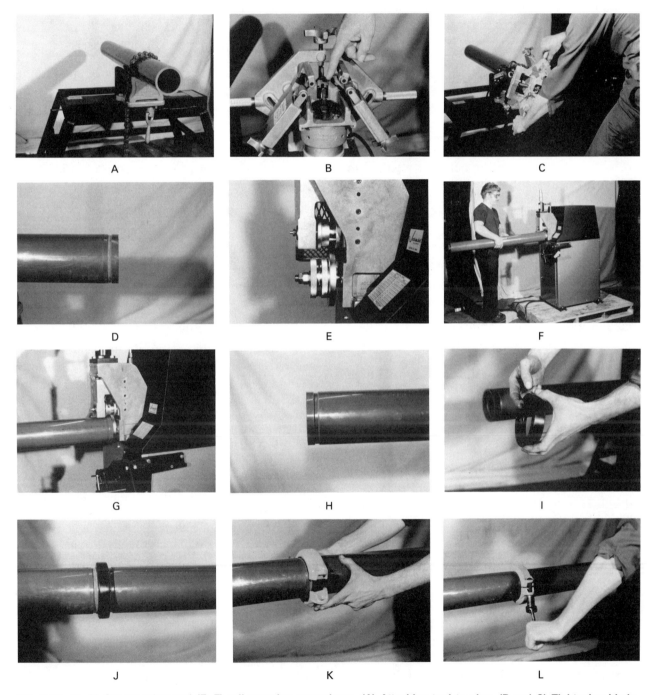

Fig. 3-16. (A–D) Cut-grooving and (E–F) roll-grooving procedures. (A) Attaching tool to pipe. (B and C) Tightening blade. (D) Cutting groove. (E) Finished product. (F) *Left*, roll grooving tool forms groove in pipe while maintaining the required dimensions; *middle*, roll grooving does not remove any material from the pipe; *right*, inner dimple does not create any appreciable pressure drop. (I–L) Examples of the compression joining. (Courtesy of Victaulic Company of America.)

tic to plastic fittings. Use Teflon tape, Teflon lubricant, or Rectorseal.®

GROOVING AND MECHANICAL JOINTS

The grooved-end mechanical joint is an excellent joining system to use when temporary piping systems are required or when environmental conditions make it very difficult to use other available joining methods. It is a proven system that has over 50 years of installation experience.

This joining system uses a mechanical coupling that secures two pipe ends together by means of a small groove around the circumference of the pipe near the pipe end. The coupling is comprised of two identical semicircular halves, which fit the grooves on the ends of each pipe section to be joined. The coupling halves are bolted together to enclose and compress an elastomer gasket. The gasket has a "C" shaped cross-section, which is pressure responsive to provide a tight seal between the gasket and the outside diameter of the pipe. A variety of gasket materials is available to ensure proper compatibility with most process fluids.

There are two methods of grooving the pipe: cut-grooving and roll-grooving. The cut-grooving method uses a pipe groover, which actually notches the pipe using a ratchet-held, spring-loaded knife (Figs. 3-16A–3-16D). This method of grooving reduces the wall thickness substantially and, consequently, reduces the working pressure of the plastic pipe. The preferred method of grooving plastic pipe is cold roll grooving.

To roll groove a pipe, a special stationary tool is used, which produces a groove to the correct dimensions by displacing the material (Fig. 3-16F). The advantage of this method is that there is no thinning of the pipe wall and no pressure derating of the pipe. Roll grooving is much quicker than cut grooving and imposes less stresses in the pipe; the stress is due to the "notching" effect of the cut-groove method. Figures 3-16F and 3-16G and 3-16H depict a rolled groove being made. The 2-in. PVC pipe groove shown was made in 5.5 sec. Figures 3-16I and 3-16L show the installation of a typical mechanical coupling on PVC 80 roll-grooved pipe.

Cut and Roll Grooving—Points to Remember

1. Not all plastic pipe lends itself to this method of joining. The polyolefins (polypropylene, polyethylene, and polybutylene) have material memory characteristics that prevent the groove from remaining dimensionally stable. Also, this method is not recommended for glass-reinforced thermosetting pipe.

2. Before grooving, check the outside pipe diameter to see if it is within tolerances. Oversized or undersized pipe will affect the groove diameter.

3. Make certain that pipe ends are cut square.

4. Use pipe supports whenever possible in grooving pipe.

5. Maintain tools for cleanliness and sharpness.

6. Use rolled grooves whenever possible. It provides a stronger system with plastic piping systems as compared to cut grooving.

Cut-Grooving—Points to Remember

1. Before grooving, check the outside pipe diameter to see if it is within tolerances. Oversized or undersized pipe will affect the groove diameter.

2. Make certain that pipe ends are cut square.

3. When grooving many pieces, a power-operated groover is available, which may be more practical.

GROOVING AND COMPRESSION (MECHANICAL)

This type of joint is made by grooving the outside diameter of a pipe. An elastomer ring is slipped over the pipe and positioned in the groove. A fitting is then placed over the pipe diameter and a compression nut is tightened to secure the pipe to the fitting. This joint is used only for Labline® polypropylene acid waste drainage systems.

Plastic Systems Applicable—For fire-retardant, glass-filled, polypropylene acid-waste drainage systems.

Instructions

1. Prepare O-ring (commonly called "olive") by immersing in hot water to soften and expand (Fig. 3-17A).

2. Cut groove in pipe by rotating tool counterclockwise as in Fig. 3-17B. Do the cutting in

two stages: (1) Adjust grooving tool for half-depth, turn, then, (2) adjust for full-depth and turn again. *Retract cutting blade* before attempting to remove tool from the pipe (Fig. 3-17C).

3. Place compression nut on pipe, then push on warm "olive" until it clicks into the groove, as shown in Fig. 3-17D. Insert this pipe end with the olive into the fitting and tighten by hand (Fig. 3-17E).

4. Further tighten nut one-quarter turn past hand-tight, using a spanner wrench (Fig. 3-17F).

Grooving and Compression—Points to Remember

1. The hot water used to heat "olives" should be between 140 and 200° F (60 and 93° C). The $1\frac{1}{2}$ and 2-in. olives require at least 10 sec and the 3- and 4-in. olives require at least 60 sec of immersion in the hot water.

2. A 4-quart pan and an inexpensive appliance equipped with a thermostat control provides a source of hot water and a means of keeping a quantity of olives ready for immediate use—thus, joints are consecutively made without waiting or interruption.

TRANSITION FROM PLASTIC TO OTHER MATERIALS

The most common methods for tying plastic pipe into other materials are through the use of flanges or threaded connections. Practically all plastic piping systems have 150 psi flanges, threaded couplings, threaded nipples, or female and male threaded adapters.

Owing to the notch sensitivity of plastics and the tendency to overtighten plastic threaded fittings, solvent or fusion-welded systems should be used wherever possible with plastic piping systems.

Another transition is plastic DWV and cast-iron pipe. The following instructions and illustrations demonstrate this type of transition joint.

Plastic/Cast-Iron Hub Pipe Instructions

1. With the end of the plastic adapter firmly seated in the cast-iron hub, lay in and pack oakum, compacting moderately with a calking iron. The oakum should only be packed about one-half as hard as when making a cast-iron joint. The oakum layer should be built up to within $1-1\frac{1}{2}$ of the top of the hub (Figs. 3-18A and 3-18B).

2. Apply a rope of lead wool all around the joint and "iron" just enough to make the lead wool

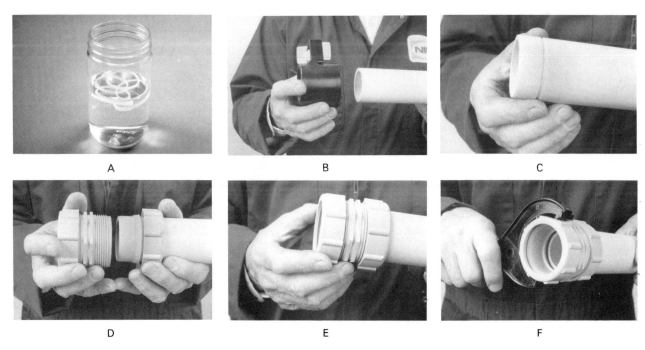

A   B   C

D   E   F

Fig. 3-17. Steps in joining acid-waste-drainage O-ring pipe and fittings.

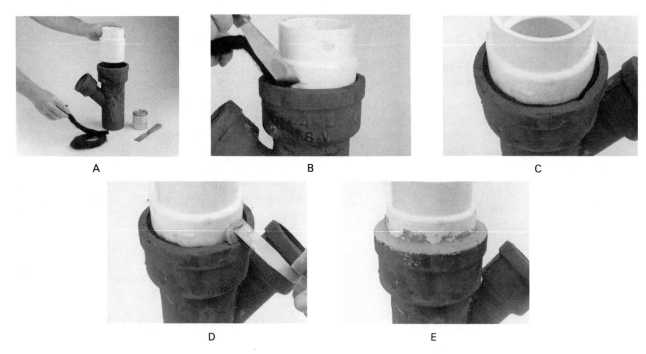

Fig. 3-18. Steps in making a cast-iron to PVC transition joint.

compact as seen in Fig. 3-18C. Excessive pressure (ironing) will form the PVC pipe out of round and allow leaks. Apply liquid lead compound with spatula until a smooth buildup is finished. Let joint cure for a minimum of 1 hr before moving. (See Figs. 3-18D and 3-18E.)

TRANSITION UNIONS

Another innovation in joining plastic piping to metal piping is the transition union (see Figs. 3-19A and 3-19B). One side of the plastic union houses a plastic socket or threaded end connector, while the opposite side (tail-piece) has a metal male or female threaded end. This product allows the varying rates of expansion or contraction of both materials to take place without any leakage. Another advantage of this adapter is that it prevents metal male threads from being overtightened in a female plastic thread connection, causing leakage or possible cracking of the plastic fitting.

Fig. 3-19. (A) PVC threaded and stainless steel male pipe thread transition union. (B) PVC socket and brass female pipe thread transition union. (Courtesy of Chemtrol Div. of Nibco.)

4

Product Selection

INTRODUCTION

To select the correct product, one must have a thorough knowledge of the application, the environment of the application, and the product. The products shown and discussed in this chapter are readily available and are, for the most part, regularly stocked by suppliers. Other types of plastic pipe, fittings, and valves are manufactured, and the supplier's or manufacturer's catalogs should be consulted for these. When selecting a product, be sure, especially if you are a neophyte in the use of plastics, that you list with the supplier the working temperature (maximum and minimum), working pressure (maximum and minimum), type of fluid to be used, its specific gravity, its viscosity, indoor or outdoor use, and any other pertinent information. The disclosure of this information will help in the selection of the proper product.

PRODUCT INFORMATION AND AVAILABILITY

PIPE

In the last 20 years, plastic pipe has become the dominant material in many markets. The gas transmission, plumbing, electrical conduit, acid waste drainage, water lines, underground irrigation, and other markets have succumbed to the many cost-saving benefits of plastic piping. Unlike the metals market, almost all domestically used plastic piping is supplied by U.S. or Cana-

dian manufacturers to standards set by such agencies as ASTM, CSA, and NSF. Color does not affect the performance of plastic pipe, and it is produced in many colors, depending on the manufacturer's or customer's preference. Most underground PVC pipe is white or light-colored to reflect sunlight more readily when waiting for burial, resulting in less warpage of pipe in outdoor installations in warm weather. Also, manufacturers can supply pipe with a large diversity of special ends. For example, listed below are just some pipe ends available in the plastics industry:

Plain × Plain

Plain × Bell

Plain × Coupling (socket or threaded)

Plain × Thread (male pipe thread)

Plain × Taper

Taper × Taper

Thread × Thread (MPT × MPT)

Plain × Grooved

Grooved × Grooved

Flanged × Flanged

All plastic pipe, with the exception of drainage pipe, is pressure rated. Pressure ratings are designated by a Schedule or a pressure rating. Schedule plastic pipe is available in Schedules 40 and 80, and, to a limited extent, 120. (The higher the Schedule number, the thicker the pipe wall.) Schedule numbering is a practice that was carried over from metal to plastic piping.

A more sensible method of classifying pipe is by a pressure rating at a given temperature. Thermosets are normally rated at a maximum temperature; thermoplastics are normally rated at ambient temperature [73–74° F (23° C)]. There are many pressure ratings commercially available, and, in fact, a manufacturer could make an infinite variety of pressure-rated pipe. The more popular thermoplastic ratings are 315 psi, 250 psi, 200 psi, 160 psi, 125 psi, 100 psi, and 50 psi.

The thermoplastic piping industry also uses a *standard dimension ratio* (SDR) to classify the pressure of pipe. The SDR ratio is determined by dividing the outside diameter of the pipe by its wall thickness. Table 4-1 lists the most common SDR's and comparable pressure ratings for plastic pipe. Thermoplastic pipe is extruded, which lends itself to excellent production rates, very smooth inside and outside surfaces, and uniformity in production. Extruded pipe is available in varying lengths and types of ends depending on the manufacturer. Most pipe is available in 10 and 20 ft lengths, but can easily be manufactured in lengths up to 60 ft. The determining length factors are volume of pipe required and mode of transportation.

In the past most thermoplastic pipe was designated by a code that consisted of four digits and a product alphabetical prefix indicating the resin. The four digits stood for:

1st digit = Type of resin

2nd digit = Grade of resin

3rd and 4th digits = Hydrostatic pressure
divided by 100

For example, CPVC 4120 meant the resin was CPVC Type 4, Grade 1 with a 2000-psi hydrostatic test pressure.

The plastics industry, with ASTM's assistance, has started to move away from a four-digit-code numbering system and into a more resin-descriptive coding called cell classification. Cell classi-

fications more quantitatively describe a resin's physical properties, eliminating confusion in the industry and in the marketplace. Listed in Fig. 4-1 is a sample cell classification for a PVC pipe resin.

ASTM standard specifications for thermoplastics list class requirements for several compounds. Listed in Tables 4-2 and 4-3 are the class requirements for rigid PVC and CPVC compounds. For further classification information, Table 4-4 lists the ASTM material code that will give in-depth coding classifications for each plastic resin.

RTRP products are filament wound, centrifu-

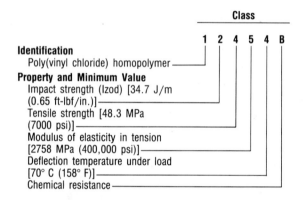

Fig. 4-1. Sample cell classification for PVC pipe resin.

Table 4-2. Class Requirements for PVC/CPVC Compounds

Designation Order Number	Property and Unit	Cell Limits	
		1	2
1	Base resin	polyvinyl chloride homopolymer	chlorinated polyvinyl chloride
2	Impact strength (Izod) minimum:		
	J/m of notch	<34.7	34.7
	ft-lb/in. of notch	<0.65	0.65
3	Tensile strength, minimum:		
	MPa	<34.5	34.5
	psi	<5000	5000
4	Modulus of elasticity in tension, minimum:		
	MPa	<1930	1930
	psi	<280,000	280,000
5	Deflection temperature under load, minimum: 1.82 MPa (264 psi):		
	°C	<55	55
	°F	<131	131
	Flammability	A[a]	A[a]

[a]A = average extent of burning 25 mm; average time of burning 10 sec (testing method 635).

Table 4-1. Comparisons of SDR Pipe to Pressure Ratings

Standard Dimension Ratio (SDR) of Pipe	Pressure Rating (psi)
13.5	315
17	250
21	200
26	160
32.5	125
41	100
64	63

Table 4-3. Suffix Designation for Chemical Resistance

Solution	A	B	C	D
H_2SO_4 (93%)—14 days immersion at 55 ± 2° C (131 ± 3.6° F):				
Change in weight:				
increase, maximum (%)	1.0	5.0	25.0	NA[a]
decrease, maximum (%)	0.1	0.1	0.1	NA
Change in flexural yield strength:				
increase, maximum (%)	5.0	5.0	5.0	NA
decrease, maximum (%)	5.0	25.0	50.0	NA
H_2SO_4 (80%)—30 days immersion at 60 ± 2° C (140 ± 3.6° F):				
Change in weight:				
increase, maximum (%)	NA	NA	5.0	15.0
decrease, maximum (%)	NA	NA	5.0	0.1
Change in flexural yield strength:				
increase, maximum (%)	NA	NA	15.0	25.0
decrease, maximum (%)	NA	NA	15.0	25.0
ASTM Oil No. 3—30 days immersion at 23° C (73.4° F):				
Change in weight:				
increase, maximum (%)	0.5	1.0	1.0	10.0
decrease, maximum (%)	0.5	1.0	1.0	0.1

[a]NA = not applicable.

Table 4-4. Pipe and Pipe-Liner ASTM Material Designation Code Reference

Pipe Material	ASTM Material Specification
PVC	D1784
CPVC	D1784
ABS	D1788
PE	D1248
PB	D2581
PP	D4101
PVDF	D3222
PFA (liner)	D3307
PTFE (liner)	D1457
FEP (liner)	D2116
RTRP	D2310

Table 4-6. Hydrostatic Design Basis Categories

	Cyclic Test Method		Static Test Method
Designation	Hoop Stress, psi (MPa)	Designation	Hoop Stress, psi (MPa)
A	2,500 (17.2)	Q	5,000 (34.5)
B	3,150 (21.7)	R	6,300 (43.4)
C	4,000 (27.6)	S	8,000 (55.2)
D	5,000 (34.5)	T	10,000 (68.9)
E	6,300 (43.4)	U	12,500 (86.2)
F	8,000 (55.2)	W	16,000 (110)
G	10,000 (68.9)	X	20,000 (138)
H	12,500 (86.2)	Y	25,000 (172)
		Z	31,500 (217)

Source: ASTM D2310.

gally cast, or pressure laminated (contact-molded or hand laid-up). Thermoset pipe manufacturing techniques are becoming less artistic, more scientific, and, recently, have had ASTM standards established. There are now code designations for RTRP as shown in Tables 4-5, 4-6, 4-7, and 4-8, which greatly facilitates the trend to more uniform standards among pipe manufacturers. Listed in Fig. 4-2 is a typical sample code designation for an RTRP filament-wound product.

Much longer cycle times are required to produce RTRP pipe as compared to thermoplastic pipe. Glass composition, resin content, wall thickness, ribbing, wall smoothness, joints, and color vary with each manufacturer. Table 4-9 shows available product and size range of RTR and thermoplastic pipe.

FITTINGS

Thermoset and thermoplastic pipe fittings are either molded or fabricated out of pipe. Thermoplastic fittings are injection molded and RTR fittings are normally compression, transfer, or contact molded. Molded fittings are usually less costly, have higher pressure ratings, and have smoother

Table 4-5. Reinforced Thermosetting Resin Pipe Designation Codes

Types	Grades	Classes[a]	Hydrostatic Resin Basis	Mechanical Properties
I Filament wound II Centrifugally cast III Pressure laminated	1 Glass-fiber reinforced epoxy 2 Glass-fiber polyester 3 Glass-fiber phenolic 4 Asbestos-reinforced polyester 5 Asbestos-reinforced epoxy 6 Asbestos-reinforced phenolic 7 Glass-fiber-reinforced Furon	A No liner B Polyester liner (NR) C Epoxy liner (NR) D Phenolic liner (NR) E Polyester liner (R) F Epoxy liner (R) G Phenolic liner (R) H Thermoplastic liner (specify) I Furon liner (R)	Hoop stress category by cyclic or static test is depicted by a letter (see Table 4-6). In filament-wound pipe the pipe is subjected to axial loading. Type of closures for this testing is done by a free-end-type closure shown by the Number 1 and a restrained end test designated by the Number 2.	See Tables 4-7 and 4-8 numeric designations.

[a]NR = nonreinforced; R = reinforced.

Table 4-7. Physical Property Requirements of Filament-Wound Pipe

Designation Order Number	Mechanical Property	Cell Limits					
		1	2	3	4	5	6
1	Short-term rupture strength hoop tensile stress, minimum, psi (MPa)	10,000 (68.9)	30,000 (207)	40,000 (276)	50,000 (345)	60,000 (414)	70,000 (483)
2	Longitudinal tensile strength minimum, psi (MPa)	8,000 (55.2)	15,000 (103)	25,000 (172)	35,000 (241)	45,000 (310)	55,000 (379)
3	Longitudinal tensile modulus, minimum, psi $\times 10^6$ (MPa)	1 (6,900)	2 (13,800)	3 (20,700)	4 (27,600)	5 (34,500)	6 (41,400)
4	Apparent stiffness factor at 5% deflection, min, in.^{3}-lbf/in.^{2} (mm^3-kPa)	40 (4.5)	200 (22.6)	1,000 (113)	1,500 (170)	2,000 (226)	2,500 (282)

Adapted from ASTM D2996.

Table 4-8. Physical Property Requirements of Centrifugally Cast Pipe

Designation Order Number	Mechanical Property	Cell Limits			
		1	2	3	4
1	Short time rupture strength hoop tensile stress, minimum, psi (MPa)	4,000 (27.6)	12,000 (82.7)	22,000 (152)	30,000 (207)
2	Longitudinal tensile strength, minimum, psi (MPa)	2,000 (13.8)	8,000 (55.2)	16,000 (110)	22,000 (152)
3	Longitudinal tensile modulus, minimum, psi $\times 10^6$ (MPa)	0.6 (4,100)	1.3 (9,000)	1.5 (10,300)	1.9 (13,100)
4	Apparent stiffness factor at 5% deflection, minimum, in.^{3}-lbf/in.^{2}	100	1,000	3,000	5,000

Adapted from ASTM D2997.

flow characteristics than fabricated fittings. The manufacturing cycle times are also much better in molded fittings than fabricated ones owing to the labor-intensive nature of the fabrication process. Most plastic fittings are molded up to 8-in. size. For sizes 10 in. and above, fittings are usually fabricated. Pressure ratings of thermoplastic fabricated fittings, if hand welded and not banded or fiberglass wrapped, are 20 psi or less only. The

use of new sophisticated butt fusion techniques or overwrapping of thermoplastic fabrications with fiberglass will allow the fittings to have similar pressure ratings as the pipe it is joining. Molded fitting resin in most cases is similar to pipe resin in physical and chemical characteristics. The pressure capabilities of molded fittings almost always exceed the joining pipe burst pressures and usually are rated similarly to pipe such as Schedule 80, Schedule 40, constant pressure, and drainage. Except for particular compression fittings, most plastic fittings are available in socket, threaded, spigot, or flanged ends. Most thermoplastic PVC, CPVC, and ABS fittings are interchangeable among manufacturers. However, thermoset fittings rarely can be interchanged with another manufacturer's product. Table 4-10 shows product sizes available and Fig. 4-3 shows several pictures of current fittings.

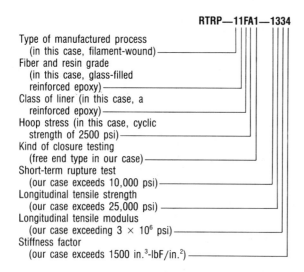

RTRP—11FA1—1334

Type of manufactured process (in this case, filament-wound)
Fiber and resin grade (in this case, glass-filled reinforced epoxy)
Class of liner (in this case, a reinforced epoxy)
Hoop stress (in this case, cyclic strength of 2500 psi)
Kind of closure testing (free end type in our case)
Short-term rupture test (our case exceeds 10,000 psi)
Longitudinal tensile strength (our case exceeds 25,000 psi)
Longitudinal tensile modulus (our case exceeding 3 $\times 10^6$ psi)
Stiffness factor (our case exceeds 1500 in.^{3}-lbF/in.^{2})

Fig. 4-2. Typical sample code designation for an RTRP filament-wound product.

VALVES

Valves control fluids in three ways:

1. Shut-off (on/off).

2. Throttling (regulation or modulation).

3. Nonreturn (prevention of reverse flow).

Table 4-9. Plastic Pipe Availability

Material	Schedule/Pressure	Nominal Size Range (in./IPS[c])	Color[a]	Lengths and Type Ends[b]	Comments
ABS	Schedule 40	1¼–6	Black	20 ft/plain end	Used in DWV applications
ABS	230 psi	1–8	Blue/gray	20 ft/plain end	Used in industrial applications
ABS	180 psi	1–6	Blue/gray	20 ft/plain end	Used in industrial applications
ABS	145 psi	⅜–4	Blue/gray	20 ft/plain end	Used in industrial applications
ABS	Airline (185 psi)	⅜–4	Light Blue	20 ft/plain end	Only plastic piping system designed to handle aboveground compressed-air lines
CPVC	Schedule 80	¼–10	Purplish gray	20 ft/plain end	Used in industrial applications over 100° F (38° C) and for threaded systems
CPVC	Schedule 40	¼–10	Purplish gray	20 ft/plain end	Used in service systems above 100° F (38° C), not used much
CPVC	Copper tube size (100 psi)	½–2	Beige	20 ft/plain end	Used in plumbing hot and cold water systems
CPVC	SDR 13.5	½–3	Orange	10 ft/20 ft plain end	For fire sprinkler systems; FM & UL approved
Polybutylene	SDR 9 (250 psi)	2–8	Black		
Polybutylene	SDR 11 (200 psi)	2–18	Black		
Polybutylene	SDR 135 (160 psi)	2–22	Black	<3 in. coiled	Used in plumbing for hot and cold water as well as industrial applications; excellent for
Polybutylene	SDR 17 (125 psi)	4–28	Black	>3 in. 20/40 ft lengths, plain ends	handling slurries; a UL and FM approved fire
Polybutylene	SDR 21 (100 psi)	4–32	Black		sprinkler system is available; also available in
Polybutylene	SDR 26 (80 psi)	4–32	Black		blue
Polybutylene	SDR 343 (60 psi)	4–36	Black		
Polybutylene	SDR 455 (45 psi)	4–36	Black		
Polyethylene	SDR 7.3 (250 psi)	2–16	Black		
Polyethylene	SDR 9 (200 psi)	¾–20	Black		
Polyethylene	SDR 11 (160 psi)	¾–24	Black		
Polyethylene	SDR 13.5 (130 psi)	2–28	Black		
Polyethylene	SDR 15.5 (110 psi)	2–32	Black		
Polyethylene	SDR 17 (100 psi)	½–36	Black	<3 in. coiled	Used in underground irrigation, gas transmission, water transmission, and
Polyethylene	SDR 21 (80 psi)	¾–42	Black	>3 in. 20/40 ft lengths, plain ends	industrial applications; the premier slip-lining
Polyethylene	SDR 26 (65 psi)	3–48	Black		plastic piping material; excellent, for handling
Polyethylene	SDR 32.5 (50 psi)	3–48	Black	Gasket pipe available from 18 to 60 in.	abrasive solutions; gas pipe is orange or
Polyethylene	Class 40	27–120 (ID)	Black		beige
Polyethylene	Class 63	21–120 (ID)	Black		
Polyethylene	Class 100	18–120 (ID)	Black		
Polyethylene	Class 160	18–78 (ID)	Black		
Polypropylene	Schedule 80	½–12	Black/natural	20 ft/plain end	Used in industrial applications; threaded system is not pressure rated
Polypropylene	Schedule 40	½–12	Black/light blue	20 ft/plain end	Used for drainage; a fire retardant grade is available
Polypropylene	150 psi	½–18	White	16 ft 4 in./plain end	Used in industrial applications and for pure water systems; not suited for use outdoors
Polypropylene	45 psi	2–24	White	16 ft 4 in./plain end	Used in industrial applications and for pure water systems; not suited for use outdoors
PVC	Schedule 120	½–6	Gray	20 ft/plain end	Normally not stocked—used infrequently
PVC	Schedule 80	⅛–16	Gray	20 ft/plain end	Used in many industrial applications; Schedule 80 may be threaded
PVC	Schedule 40	⅛–16	White	20 ft/plain end	Used for pressure and DWV
PVC	SDR 13.5 (315 psi)	½	White	20 ft/plain end	Used infrequently
PVC	SDR 17 (250 psi)	¾–6	White	20 ft/plain end	Used infrequently
PVC	SDR 21 (200 psi)	¾–16	White	20 ft/plain end and gasket	Used for all types of below ground water applications
PVC	SDR 26 (160 psi)	1–20	White	20 ft/plain end and gasket	Used for all types of below ground water applications
PVC	SDR 32.5 (125 psi)	2–24	White	20 ft/plain end	Used infrequently

Table 4-9. (*Concluded*). Plastic Pipe Availability

Material	Schedule/Pressure	Nominal Size Range (in./IPS[c])	Color[a]	Lengths and Type Ends[b]	Comments
PVC	SDR 35	4–27	White	20 ft/plain end and gasket	Used in sewer applications
PVC	SDR 41 (100 psi)	3–24	White	20 ft/plain end	Used mostly in below ground irrigation and drainage applications
PVC	SDR 100 (50 ft Head)	6–15	White	20 ft/plain end	Used in drainage applications
PVC	Sewer/Drain	3–6	White	10 and 20 ft lengths plain end	Available in perforated and nonperforated; used for drains
PVC	C900 (AWWA Type) Class 100/DR25 Class 150/DR18 Class 200/DR14	4–12 4–12 4–12 4–12	White/blue/green	12.5 ft lengths, gasketed	Used for water lines (both main and transmission); pipe size is cast iron dimensions
PVDF	Schedule 80	1/2–4	Red/natural	20 ft plain ends	Used in very corrosive environments and for threaded and socket fusion applications; also used in pure water and severe halogen service applications
PVDF	232 psi	3/8–4	Natural/white	16 ft 4 in. plain ends	
PVDF	150 psi	3–12	White	16 ft 4 in. plain ends	
RTRP	150 psi Epoxy (lined and unlined)	1/2–144	Varies	20 ft nominal lengths, plain, tapered, bell spigot, and gasketed, threaded ends available	Used in all types of industrial, municipal, commercial, and military, applications; pressures vary from 25 to 300 psi depending on the size, the manufacturer, and the product type; the limiting diameter size of any RTRP product is the method of shipment
RTRP	150 psi polyester (lined and unlined)	1/2–144	Varies		
RTRP	150 psi vinylester (lined and unlined)	1/2–144	Varies		

[a]The colors listed are the most commonly used. Depending on the volume, a manufacturer could produce pipe in any desired color and, in some cases, even clear or transparent. [b]The listed lengths and type pipe ends are the most common. However, most manufacturers could easily produce the pipe in any desired length (method of pipe transportation is limiting factor) and in various ends such as threaded, flanged, grooved, tapered, etc. [c]IPS = iron pipe size.

Table 4-10. Commonly Available Plastic Shutoff Valves

Type	Material[a]	Size Range (in.)	Color	Types of Ends[b]
Ball	PVC	1/4–4	Gray	Soc/Thd/Flg/Spig
	CPVC	3/8–4	Purplish gray	Soc/Thd/Flg/Spig
	Polypropylene	3/8–4	Black/natural	Soc/Thd/Flg/Spig
	PVDF	1/2–4	Red/natural	Soc/Thd/Flg/Spig
	FRP	1/2–8	Gray/white	Flanged
	PPS[c]	1–4	Black	Flanged
	ABS	1/2–3	Bluish gray	Soc/Spig/Flg
Gate	PVC	1/2–14	Blue	Soc/Thd/Flg
	PVDF	11/2–6	Natural	Flanged
	Celcon®	1/2–2	White/dark gold	Threaded
Multiport	PVC	1/2–6	Gray	Soc/Thd/Flg
	CPVC	1/2–6	Purplish gray	Soc/Thd/Flg
	PVDF	1/2–4	Red/natural	Soc/Thd/Flg
	ABS	1/2–2	Bluish gray	Soc/Flg
Solenoid	PVC	1/8–3	Gray	1/8–1/2 in.—Soc/Thd Above 1/2 in.—Threaded
	CPVC	1/8–3	Purplish gray	1/8–1/2 in.—Soc/Thd Above 1/2 in.—Threaded
	Polypropylene	3/4–3	Natural	1/8–1/2 in.—Soc/Thd Above 1/2 in.—Threaded
Relief	PVC	1/8–2	Gray	Threaded
	CPVC	1/2–2	Purplish gray	Threaded
Butterfly	PVC	11/2–24	Gray	Flanged
	Polypropylene	11/2–24	Natural	Flanged
	PVDF	11/2–24	Natural	Flanged
	FRP	2–12	Blue/black	Flanged

[a]Celcon® = Celanese trademark.
[b]Soc = socket; Thd = threaded; Flg = flanged; Spig = spigot.
[c]PPS = polyphenylene sulfide.

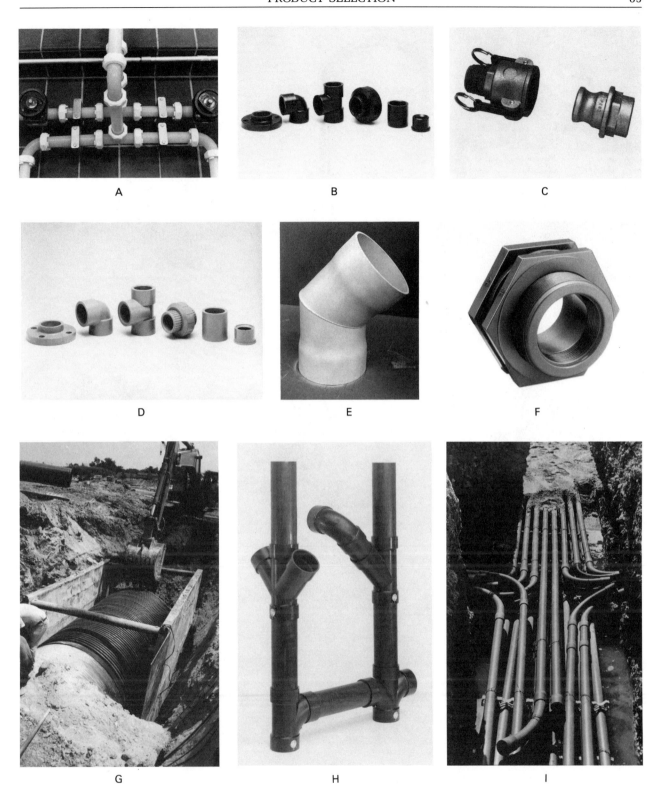

Fig. 4-3. Piping systems: (A) 2-in. PVDF sanitary mechanically joined piping system (courtesy of Sani-Tech Inc.); (B) polypropylene fittings (courtesy of Chemtrol Div. of Nibco); (C) 2-in. glass-filled polypropylene quick-disconnect fittings (courtesy of Chemtrol Div. of Nibco); (D) CPVC fittings (courtesy of Chemtrol Div. of Nibco); (E) 16-in. PVC fabricated 45° ell (courtesy of Plastinetics); (F) 2-in. PVC bulkhead (tank adapter) fitting (courtesy of Hayward Manufacturing); (G) 120-in. spirally extruded polyethylene pipe (courtesy of Spirolite Corp.); (H) ABS sanitary piping system (courtesy of Chemtrol Div. of Nibco); (I) PVC electrical conduit (courtesy of Chemtrol Div. of Nibco).

Fig. 4-4. Plastic valves. (A) 4-in. PVC molded basket strainer (courtesy of Hayward Manufacturing); (B) PVC laboratory goose-neck (courtesy of Plastinetics); (C) $\frac{1}{4}$-in. PVC chemcock valve (courtesy of Chemtrol Div. of Nibco); (D) 4-in. PVC gate valve (courtesy of Asahi America); (E) 2-in. CPVC diaphragm valve (courtesy of Chemtrol Div. of Nibco); (F) 6-in. PVC horizontal swing check valve (courtesy of Asahi America); (G) cross section of a 1-in. PVC relief valve (courtesy of Plast-o-matic); (H) $1\frac{1}{2}$-in. PVC globe valve (courtesy of Asahi America); (I) 1-in. PVDF true union ball valve (courtesy of Chemtrol Div. of Nibco).

Table 4-11. Commonly Available Plastic Throttling Valves

Type	Material[a]	Size Range (in.)	Color	Types of Ends[b]
Globe	PVC	$1/2$–4	Gray	$1/2$–2 in.—Soc/Thd/Flg Above 2 in.—Flanged
	Polypropylene	$1/2$–4	Natural	$1/2$–1 in.—Soc/Thd/Flg Above 1 in.—Flanged
Y-angle globe	PVC translucent	$1/2$–3	Translucent	Spig/Soc/Thd/Flg
Angle globe	PVC	$1/4$–2	Gray	Threaded
	CPVC	$1/4$–$1/2$	Purplish gray	Threaded
	Polypropylene	$1/4$–$1/2$	Natural	Threaded
	Halar®	$1/4$–$1/2$	Natural	Threaded
	PVDF	$1/4$–$1/2$	Natural	Threaded
Needle globe	PVC	$1/4$–$1/2$	Gray	Threaded
	CPVC	$1/4$–$1/2$	Purplish gray	Threaded
	Halar®	$1/4$–$1/2$	Natural	Threaded
	Polypropylene	$1/4$–$1/2$	Natural	Threaded
	PVDF	$1/4$–$1/2$	Natural	Threaded
Diaphragm	PVC	$1/2$–10	Gray	$1/2$–4 in.—Soc/Thd/Flg/Spig Above 4 in.—Flanged
	CPVC	$1/2$–10	Purplish gray	$1/2$–4 in.—Soc/Thd/Flg/Spig Above 4 in.—Flanged
	Polypropylene	$1/2$–10	Natural	$1/2$–2 in.—Soc/Thd/Flg Above 2 in.—Flanged
	PVDF	$1/2$–10	Natural	$1/2$–2 in.—Soc/Thd/Flg Above 2 in.—Flanged
	ABS	$1/2$–3	Bluish gray	$1/2$–3 in.—Soc/Flg
Labcock	PVC	$1/4$	Gray	MPTxMPT/MPTxHose MPTxFPT/FPTxFPT
Pressure regulator	PVC	$1/4$–3	Gray	Threaded

[a]Halar® = Tradename of Allied Corporation's fluorocarbon (PCTFE).
[b]Soc = socket; Thd = threaded; Flg = flanged; Spig = spigot; MPT = male pipe thread; FPT = female pipe thread.

Table 4-12. Commonly Available Plastic Nonreturn Valves

Type	Material	Size Range (in.)	Color	Types of Ends[a]
Ball check and foot	PVC	$1/2$–4	Gray	Soc/Thd/Flg
	CPVC	$1/2$–4	Purplish gray	Soc/Thd/Flg
	Polypropylene	$1/2$–4	Black/natural	Soc/Thd/Flg
	PVDF	$1/2$–4	Red/natural	Soc/Thd/Flg
	ABS	$1/2$–3	Bluish gray	Soc/Flg
Swing check	PVC	$3/4$–8	Gray	Flanged
	Polypropylene	$3/4$–8	Natural	Flanged
	PVDF	$3/4$–8	Natural	Flanged
Lift check	PVC	$1/2$–4	Gray	Soc/Thd/Flg
Y-angle check	PVC	$1/2$–4	Gray	Soc/Thd/Flg
	PVC	$1/2$–3	Translucent	Spig/Soc/Thd/Flg
Diaphragm check and foot	PVC	$1/8$–1	Gray	Threaded
	Polypropylene	$1/8$–1	Natural	Threaded
	Teflon	$1/8$–1	Natural	Threaded

[a]Soc = socket; Thd = threaded; Flg = flanged; Spig = spigot.

Table 4-13. Commonly Available Plastic Fluid Flow Products

Type	Material[a]	Size Range (in.)	Color	Types of Ends[b]
Goosenecks	PVC	¼–½	Gray	Threaded
	Polypropylene	¼–½	Natural	Threaded
	PVDF	¼–½	Natural	Threaded
Sight Glasses	PVC	½–3	Gray	Thd/Flg
	Polypropylene	½–3	Natural	Thd/Flg
Y-strainers	PVC	½–4	Gray/translucsent	Soc/Thd/Flg/Spig
	CPVC	½–4	Purplish gray	Soc/Thd/Flg/Spig
	ABS	⅜–3	Bluish gray	Soc/Flg
Basket strainers	PVC	½–4	Gray	Soc/Thd/Flg
	CPVC	½–4	Purplish gray	Soc/Thd/Flg
	FRP	3–12	White	Flanged
	Polypropylene	½–8	Natural	Soc/Thd/Flg to 4 in. Above 4 in. Flanged
Float valve	PVC	¼–1	Gray	Threaded
	Polypropylene	¼–1	Natural	Threaded
Gage guard	PVC	¼–½	Gray	Threaded
	PVDF	¼–½	Natural	Threaded
	Polypropylene	¼–½	Natural	Threaded
	Halar®	¼–½	Natural	Threaded
Vacuum breakers	PVC	½–1	Gray	Threaded

[a]Halar® = Allied Corp. tradename.
[b]Soc = socket; Thd = threaded; Flg = flanged; Spig = spigot.

Table 4-14. Plastic Fittings Availability

Material	Schedule/Pressure	Nominal Size Range (in./IPS)	Color[a]	Types of Ends[a]	Comments
ABS	DWV	1¼–6	Black	Socket	Molded
ABS	Air line	⅜–6	Blue	Socket	Molded
ABS	Schedule 80	⅜–8	Light gray	Socket	Molded
CPVC	Schedule 80	¼–10	Purplish/gray	Soc/Thd/Flg	Molded from ¼–8 in.; 10-in. fittings are fabricated
CPVC		½–2 (CTS)	Beige	Soc/Thd/Compression	Molded
CPVC	SDR 13.5	½–3	Orange	Socket	Molded (for fire sprinkler systems)
Polybutylene	SDR schedules	½–36	Black/blue	Spigot	Molded ½–2 in. (for sprinkler and hot–cold water systems); above 2 in. fabricated from pipe (derated fitting pressure is 80% of pipe rating)
Polyethylene	160 psi/others	¾–48	Black/orange	Soc/Spigot	Molded to 12 in.; fabricated above 12 in.
Polypropylene	Schedule 80	½–12	Black	Soc/Thd/Flg	Molded from ½ to 6 in.; above 6 in. is fabricated; threaded fittings up to 4 in. for 20 psi maximum working pressure only
Polypropylene	150 psi	½–18	White	Spigot	Molded to 6 in.; fabricated above 6 in.
Polypropylene	45 psi	½–24	White	Spigot	Molded to 6 in.; fabricated above 6 in.
Polypropylene	DWV	1¼–12	Black	Socket/Compression	Socket from 1¼ to 12 in.; compression type up to 4 in. size (for acid waste service)
PVC	Schedule 80	¼–16	Gray	Soc/Thd/Flg	Molded from ¼ to 8 in.; mostly fabricated above 8 in.
PVC	Schedule 40	⅜–16	White	Socket	Molded from ¼ to 8 in.; mostly fabricated above 8 in.
PVC	DWV	1¼–16	White	Soc/Gasket	Molded 1¼ to 8 in.; fabricated above 8 in.
PVC	200 psi/160 psi	1½–8	White	Soc/Gasket/Compression	Molded/compression fittings to 6 in.
PVC	Sewer and drain	3–6	White	Soc/Gasket	Molded
PVC	Sewer	4–27	White		Molded to 8 in.; mostly fabricated above 8 in.
PVC	Insert type	½–4	Gray	Barbed	Used with coiled tubing
PVC	200/160 psi	1½–2 IPS and CTS	White	Compression	Used with coiled tubing

Table 4-14. (*Concluded*)

Material	Schedule/Pressure	Nominal Size Range (in./IPS)	Color[a]	Types of Ends[a]	Comments
PVC	C-900	4–12	White	Gasket	Molded 4–8 in.; fabricated above
PVDF	Schedule 80	$^1/_2$–4	Red/natural	Soc/Thd/Flg	Molded 4–8 in.; fabricated above
PVDF	150 psi	3–12	Natural	Spigot	Molded to 6 in.; fabricated above 6 in.
PVDF	232 psi	$^3/_8$–4	Natural	Spigot	Molded
RTRP	150	$1^1/_2$–144	Varies	Soc/Flg/Spig	Molded to 12 in.; fabricated above 12 in.

[a]The listed colors are the most commonly used. Depending on the volume, a manufacturer could produce fittings in any desired color and, in some cases, even clear or transparent.

[b]IPS = iron pipe size; CTS = copper tube size.

[c]Soc = socket; Thd = threaded; Flg = flanged; Spig = spigot.

Plastic valves, no differently than metal valving, fall into these three categories and are also constructed with parts of similar nomenclature such as stem or shaft, seats, seals, bonnet, handwheel, and lever. The major differences in plastic versus metal valves are that plastic valves are much lighter, generally have better chemical resistance, have a little less friction loss through the valve, and lower temperature and pressure limitations. Most thermoplastic valves have a maximum op-

Fig. 4-5. Plastic valves: (A) $^3/_4$-in. PVC foot valve (courtesy of Chemtrol Div. of Nibco); (B) 6-in. PVC flanged diaphragm valve with stem indicator (courtesy of Asahi America); (C) 1-in. PVC flow indicator (courtesy of Plast-o-matic Valves); (D) $1^1/_2$-in. PVC electrically activated multiport valve (courtesy of Chemtrol Div. of Nibco); (E) 18-in. polypropylene gear-operated butterfly valve (courtesy of Asahi America); (F) $^1/_2$-in. CPVC pneumatically activated ball valve (courtesy of Chemtrol Div. of Nibco); (G) 8-in. PVC butterfly valve with extended gear-operated handle (courtesy of Chemtrol Div. of Nibco); (H) $^3/_4$-in. PVC true union solenoid valve (courtesy Hayward Manufacturing); (I) 3-in. electrically actuated butterfly valve (courtesy Chemtrol Div. of Nibco).

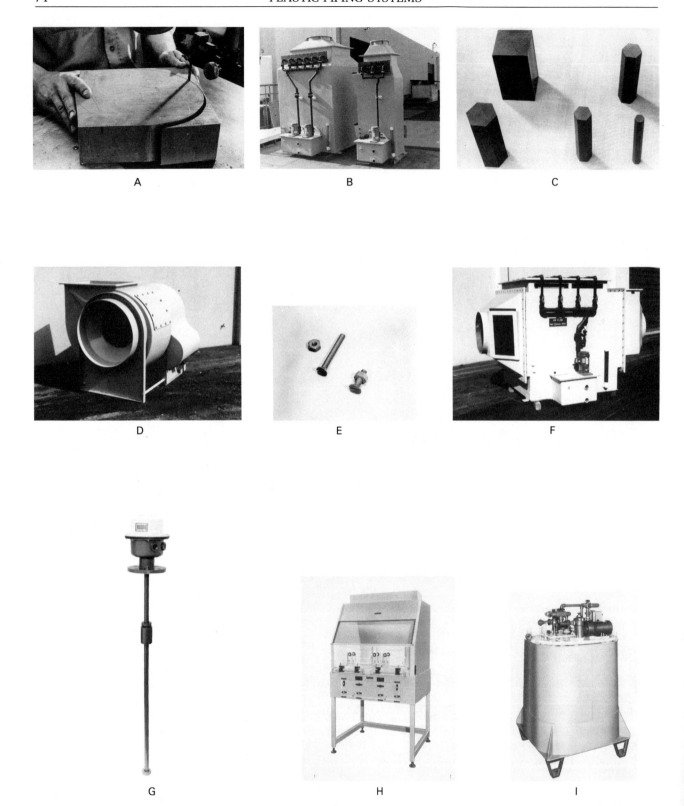

Fig. 4-6. Plastic fabrications: (A) 2-in. PVC sheet being cut (courtesy of Plastinetics); (B) fiberglass scrubber assemblies (courtesy of Harrington Plastics); (C) PVC rod and bar (courtesy of Plastinetics); (D) fiberglass blower (courtesy of Harrington Plastics); (E) PVC nuts and bolts (courtesy of Plastinetics); (F) fiberglass scrubber assembly (courtesy of Harrington Plastics); (G) all plastic liquid level controller (courtesy of Plastinetics); (H) all plastic semiconductor work station (courtesy of Plastinetics); (I) polypropylene and fiberglass holding and pumping tank neutralization systems (courtesy of Plastinetics).

erating pressure of 150 psi at 73.4° F (23° C). As the temperature increases, as discussed in Chapter 2, the pressure rating of the valve decreases. Socket, threaded, flanged, and spigot ends are available on most plastic valves. Fluorocarbon elastomeric seats and seals are most common in plastic valving with EPDM (ethylene propylene dimonomer) and Buna-N elastomers also available. There are pneumatic and electric actuators for plastic ball, multiport, diaphragm, and butterfly valves. Several plastic valve actuators have housings constructed of thermoplastics, thus making the actuator lighter in weight and more chemically resistant than metal valve actuators. The actuators are also more compact and weigh less than metal valves owing to the lower ranges of torque requirements of the plastic valves. Listed in Tables 4-10–4-14 are most of the common off-the-shelf plastic valves. Figures 4-4 and 4-5 depict several of the most commonly used valves.

FABRICATIONS

It is literally true that if you can draw a desired fluid handling product, it can be fabricated in a variety of thermoplastic and thermoset materials. Manufactured solid and hollow bar, sheet, and other profiles are available to customize almost any product. Listed in Table 4-15 are several

Table 4-15. Partial Listing of Plastic Fabrications

Pipe headers	Trays
Scrubbers	Manifolds
Fans/blowers	Dampers
Duct	Spargers
Large-diameter fittings	Air hoods
Tanks	Faucets
Sinks	Louvers
Work stations	Pails
Well screens	Float controllers

products (by no means an exhaustive list) that have been fabricated successfully. Be sure when requesting fabricated products that you specify to the fabricator the fluid or air handling environment. *List the type of fluid or gas, maximum pressure and temperature, capacity, specific gravity, static head, viscosity, percentage of solids, indoor or outdoor installation, code requirements, union or nonunion label* and other pertinent information. The prudent buyer should select only experienced fabricators who are primarily plastic specialists. Ask to see their fabrication facilities and request a reference list of customers. If you are not aware of any local fabricators, call a thermoplastic or thermoset product manufacturer; they can usually refer you to a quality and experienced fabricator. Figure 4-6 illustrates several fabricated products of interest.

5
Installing Plastic Pipe

INTRODUCTION

Reputable plastic piping manufacturers have adopted exacting product quality-control procedures in cooperation with testing agencies and professional associations, but the best-engineered product will perform as expected only if it is properly installed. Deviations from proven installation techniques can create costly repair problems that would not have existed if the system had been installed according to the manufacturer's recommendations.

When installing a system either above or below ground, with a few exceptions, plastic piping intallation practices differ very little from those of other pipe systems. However, plastic piping materials have certain unique physical characteristics that must be considered.

A basic step in dealing with plastics for the first time is to realize that plastic pipe is a new material; thus, set about learning as quickly and thoroughly as possible the accepted ways of using it properly. Plastic pipe manufacturers, plastic piping specialty distributors, or experienced contractors are good sources for obtaining the necessary information on plastic installation techniques.

INSTALLATION CREWS

Today's piping contractor normally has more money allocated for labor on a job than for material and, consequently, is more concerned with worker productivity. With a plastic material to be installed by a work crew unfamiliar with plastics, always educate and instruct the workers on the proper material handling and joining techniques for a minimum of several hours or for two to three days, depending on the difficulty and magnitude of the job. Have the crew make several practice joints. Practical instructions will increase productivity in two ways: The workers will understand how properly to install the plastic material; and they will be more self-confident.

There are cases where large users of plastics cooperate with local unions and contractors in a qualification program. In this manner, the workers can earn a qualification certificate in a short instruction program and get first preference for working on plastic piping projects.

The size of work crews for a particular installation depends on a number of variables such as size and length of pipe, the atmospheric temperature, construction conditions, construction time schedule, amount of pipe to be installed, worker's experience, the type of joining method, and local union requirements. Table 5-1 is a guide for the size of joining crews required for cementing thermoplastic and thermoset piping systems.

STORAGE AND HANDLING OF PLASTIC PIPE

STORAGE

Plastics have excellent resistance to weathering and may be stored outside for short periods of time. However, for prolonged outside storage, it is recommended that the pipe and fittings be kept under a light tarpaulin or in a well-ventilated shed to prevent excessive heat build-up and possible surface ultraviolet degradation. This will mini-

Table 5-1. Guide to Crew Size for Joining Plastic Piping Systems

Pipe Diameter (in.)	Joining Work Crew (minimum)
$1/4$–$1 1/4$	1
$1 1/2$–3	2
4–8	3
10 and above	4

mize possible warpage. Covering the piping products will keep them clean, dry, deiced, and ready for immediate installation. Many RTRP manufacturers recommend that machined joining surfaces of RTR pipe and fittings be protected from sunlight during storage to prevent possible ultraviolet degradation.

To prevent pipe diameters from extraordinary deflection, loose pipe stacks should not exceed 3 ft in height. Bundled pipe may be double-stacked. If a rack is used for pipe storage, the rack should be free from sharp metallic burrs and edges and away from heat sources such as radiators and steam pipes. Pipe racks should incorporate continuous flooring to provide good support and prevent warpage.

If belled pipe is to be stored for an extended period, it is recommended that alternate rows of bells be inverted so that loading on the bell is minimized.

HANDLING

Excessively rough handling of plastic pipe and fittings may cause damage. Dragging the free end of pipe off a truck and allowing it to crash to a hard surface can mar or split the end of the pipe. A scratch or gouge in a piece of pipe can reduce its pressure rating. It is suggested that a minimum of three people be used to unload truckloads of pipe. *Do not throw, whip, or drop the pipe during handling.*

Pipe and fittings should not contact sharp projections, be dropped, or have heavy objects dropped on them. If an accident occurs, examine the damaged section of pipe and cut it out if there is the least question as to its ability to function normally. Pads of old fire hose or rubber strips are successfully used for resting the pipe on a truck bottom. For lifting lengths of pipe, nylon or rope slings 3- or 4-in. wide should be used. Chains of any kind should not be used. *Most thermoplastic piping becomes brittle in low temperatures.* Be especially careful not to drop or impact on plastic pipe when this condition occurs.

ABOVEGROUND INSTALLATION

Good piping design techniques are recommended for all piping systems whether of plastics or other material. Anchoring, support spacing, and hanger design are somewhat unique in plastics, however, and will be discussed in this section. Another important point to consider in plastic pipe usage aboveground is that to prevent impact damage from exterior sources such as fork-lifts, hammers, and other contact devices, protective metal or wooden shields should be used.

SUPPORT SPACING

The tensile and compression strengths of plastic pipe are not as high as those of metal pipe, and, consequently, plastics require additional support. The tensile strength of thermoplastic pipe decreases as temperature increases; therefore, as the temperature increases, additional supports are required. At very elevated temperatures, continuous support is required.

Thermoset support recommendations more closely resemble metal pipe support recommendations than those for thermoplastic support. Support spacing for thermosets varies slightly from that for steel.

Table 5-2 shows the support spacing recommendations for commonly used plastic pipe.

HANGERS

The use of improper hangers can create unnecessary stresses in the pipe wall and can cause premature pipe failures. The rule-of-thumb is to use hangers that have a large bearing area so that the load is spread out over the largest practical area.

A wide variety of hangers used for metallic pipe is available. After simple modification these are also suitable for use with plastics. Figure 5-1 illustrates several types of hangers suggested for plastic piping.

For horizontal lines, consider a sling (Fig. 5-1F), clamp (Figs. 5-1C and 5-1D), or a clevis-type hanger as shown in Fig. 5-1A. A simple shoe support, shown in Fig. 5-2G, is also acceptable where other structures will permit their use. In all cases, use a hanger that lends a wide support base to the pipe. A sheet metal sleeve, under the pipe or between the pipe and the hanger (see Figs. 5-2E, 5-2F, and 5-2H), is highly desirable because it spreads the load over a broad area of the pipe and

Table 5-2. Support Spacing, in Feet,

Nominal Pipe Diameter (in.)	PVC								CPVC			
	Schedule 40				Schedule 80				Schedule 80			
	60° F (16° C)	100° F (38° C)	140° F (60° C)	180° F (82° C)	60° F (16° C)	100° F (38° C)	140° F (60° C)	180° F (82° C)	60° F (16° C)	100° F (38° C)	140° F (60° C)	180° F (82° C)
1/2	4½	4	2½	Do not use for this temperature	5	4½	2½	Do not use for this temperature	5½	5	4½	2½
3/4	5	4	2½		5½	4½			6	5½	4½	2½
1	5½	4½	2½		6	5	3		6½	6	5	3
1½	6	5	3		6½	5½	3½		7	6½	5½	3½
2	6	5	3		7	6	3½		7½	7	6	3½
3	7	6	3½		8	7	4		9	8	7	4
4	7½	6½	4		9	7½	4½		10	9	7½	4½
6	8½	7½	4½		10	9	5		11	10	9	5
8	9	8	4½		11	9½	5½		Not Available			
10	10	8½	5		12	10	6		Not Available			

ᵃFor the listed support spacings, the specific gravity of the fluid is assumed to be 1.0. For higher or lower specific gravities, adjust the listed support spacings by

reduces stresses. A U-bolt hanger or roller hanger (Fig. 5-1B) should only be used with a protective sleeve over the pipe. These sleeves can be made of medium gage sheet metal or a section of plastic pipe cut in half, axially, and fitted or "snapped" on the pipe (see Figs. 5-2E, 5-2F, and 5-2H). These sleeves will stay in place without being bonded to the pipe. If many closely spaced supports are necessary because of high temperatures, it is best to use a continuous support; a smooth structural angle, or channel, is ideal for this purpose (see Fig. 5-2A).

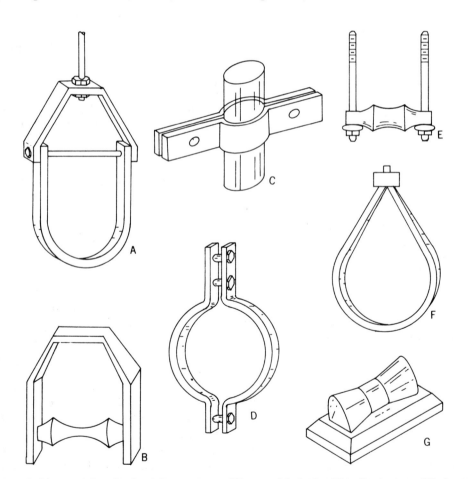

Fig. 5-1. Recommended hangers for plastic piping systems: (A) wrought clevis; (B) roller hanger; (C) riser clamp; (D) pipe roll and plate; (E) single pipe roll; (F) adjustable solid ring (swivel type); (G) double-belt pipe clamp.

for Commonly Used Plastic Pipe[a]

Polypropylene				Fiberglass Reinforced				PVDF			
Schedule 80				200 psig				Schedule 80			
60° F (16° C)	100° F (38° C)	140° F (60° C)	180° F (82° C)	60° F (16° C)	100° F (38° C)	140° F (60° C)	180° F (82° C)	80° F (27° C)	100° F (38° C)	140° F (60° C)	180° F (82° C)
4	3	Continuous support		Not Available				4½	4½	2½	Continuous support
4	3			Not Available				4½	4½	3	
4½	3			9	8½	8½	7½	5	4¾	3	
5	3½	2		9	8½	8½	7½	5½	5	3	
5	3½	2		9	8½	8½	7½	5½	5¼	3	
6	4	2½		10	9½	9½	9	Not Available			
6	4½	3		11	10½	10½	9	Not Available			
6½	5	3		12	12	11½	11	Not Available			
7	5½	3½		13½	13½	12½	11½	Not Available			
Not Available				16	15½	13½	13	Not Available			

multiplying the table values by the actual specific gravity.

In cases where the pipe may move axially because of thermal expansion due to fluid or extreme environmental temperature variations, or because of pressure changes, it may be necessary to use roller hangers as in Figs. 5-1B, 5-1E, and 5-1G. With roller hangers also, a protective shield between the pipe and the roller is both desirable and recommended. The plastic pipe must never be allowed to rub against a steel support or any other abrasive surface. For example, if the pipe rests on concrete piers, soft pine or redwood pads should be used between the pipe and the concrete surface. If weathering becomes a problem or atmospheric conditions exclude the use of wooden

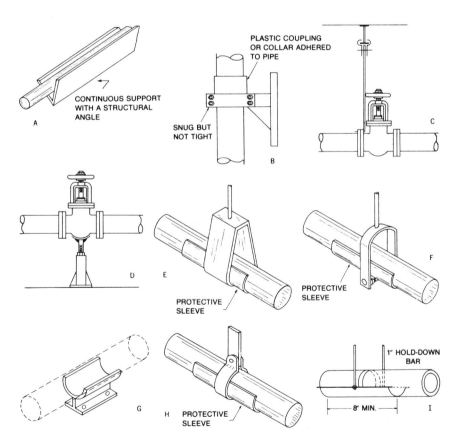

Fig. 5-2. Hangers and supports for plastic piping and valves: (A) channel support; (B) supporting plastic pipe vertically; (C) overhead support for valve; (D) valve support from below; (E) hanger with protective sleeve; (F) hanger with protective sleeve; (G) shoe support; (H) hanger with protective sleeve; (I) trapeze support.

pads, a thermoplastic pad such as PVC or polyethylene should be used. Place hangers as close as possible to elbows.

Vertical lines must be supported at intervals to avoid placing too much load on a fitting at the lower end. This can be done by using riser clamps or double-belt pipe clamps. These clamps must not exert compressive stresses on the pipe. If convenient, the clamps can be located just below a coupling so that the shoulder of the coupling rests against the clamp. If necessary, the socket portion of a fitting can be cut in half, lengthwise, and adhered to the pipe as a bearing support such that the shoulder of the fitting rests on the clamp (see Fig. 5-2B). In any event, avoid the use of riser clamps (Fig. 5-1C), which squeeze the pipe and depend on compression of the pipe to support the weight.

All plastic valves larger than 2 in. used in a piping system must be supported. In turn, the valves support the plastic pipe, rather than the pipe carrying the valves (see Figs. 5-2C and 5-2D).

Good Hanging Practice

Good plastic piping hanging practice may be summed up as follows:

1. Avoid point contact or concentrated bearing loads on small areas of pipe.

2. Avoid abrasive contact.

3. Use protective shields to spread the loads over large areas.

4. Do not have the pipe support heavy valves or specialty fittings.

5. Do not use hangers that "squeeze" the pipe.

ANCHORS AND GUIDES

Anchors in a piping system direct movement of pipe within a defined reference frame. At the anchoring point, there is no axial or transverse movement. Guides are used to allow axial movement of pipe but prevent transverse movement. Anchoring and guides should be engineered to provide the required function without point loading the plastic pipe. Some typical anchors for plastic pipe are illustrated in Fig. 5-3.

Guides and anchors are used whenever expansion joints are utilized and are also used on long runs and directional changes in piping. Anchors should be placed as close to an elbow as possible.

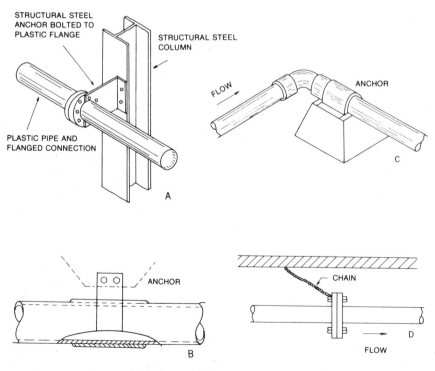

Fig. 5-3. Methods of anchoring plastic pipe: (A) plastic pipe with metal anchor; (B) plastic pipe with metal sleeve and anchor; (C) plastic pipe with concrete anchor; (D) plastic pipe with metal chain anchor.

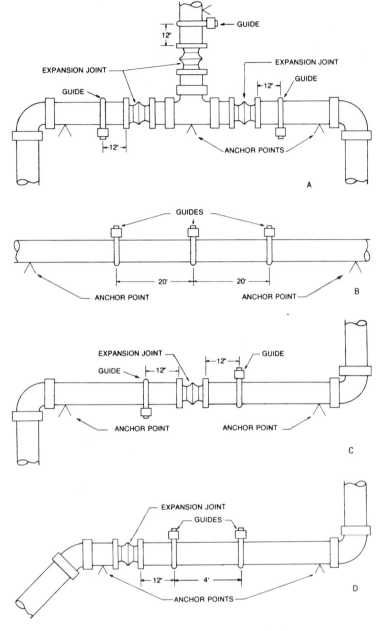

Fig. 5-4. Anchoring and guiding plastic pipe.

Illustrated in Fig. 5-4 are examples of proper anchoring and use of guides with plastic pipe.

ABOVEGROUND INSTALLATION OF PLASTIC PIPE—POINTS TO REMEMBER

1. The installation crew must be knowledgeable in both the handling of plastic material and the required joining method.

2. Be sure to handle and store plastic pipe carefully and sensibly.

3. Do not forget expansion or contraction considerations when designing for a piping system.

4. Determine the required support spacing and use properly designed hangers.

5. If anchors or guides are required, use them as suggested in the text.

6. Do not use plastic pipe aboveground to support valves, pumps, etc.

7. Protect plastic pipe with a collar of wood or steel if it may be damaged by impact with forklifts or other devices.

8. Keep hangers as close to elbows as possible.

9. Place anchors as close to elbows as possible.

10. Where many direction changes occur in a piping system, anchor the system at alternate corners.

BELOW-GROUND INSTALLATION

Because of its light weight, its corrosion resistance to all types of soil, and its easy joining methods, plastics are ideal for below-ground use.

The pipe should be stored and handled carefully, as previously described. In stringing out the pipe, these points should be kept in mind:

1. The socket ends should point in the direction the work is progressing.

2. If the trench has not yet been dug, string the pipe some distance away from the excavation equipment.

3. If the trench has been dug, string the pipe as close to the trench as possible.

4. If vandalism is a hazard, string only enough pipe for one day's use.

TRENCHING

The trench should be as narrow as possible, yet wide enough to allow the convenient installation of plastic pipe. Minimum suggested trench widths are listed in Tables 2-7 and 2-8. A rule of thumb is to have a trench width that is less than three times the pipe diameter. A narrow trench allows the pipe to be joined and tested aboveground, and then laid in the trench.

GRADING/BEDDING

The trench bottom should be continuous, relatively smooth, and free of rocks. Where ledge rock, hardpan, or boulders are encountered, it is advisable to pad the trench bottom using a minimum of 4–6 in. of tamped earth or sand as a cushion to protect the pipe from damage. For pressure systems, such as watermain or transmis-

sion lines, accurate trench leveling is not essential. For gravity piping systems, the trench bottoms should be evenly graded, as it is for other piping materials.

PLACEMENT IN THE TRENCH

Most plastic pipe sizes up to 8-in. diameter can usually be installed manually into the trenches. Most plastic pipe can snake and bend from the trench surface to the bed of the trench. With larger diameter pipe, some equipment needs to be used. Typical equipment includes backhoes, cranes, and telescoping lifting equipment. For lifting long joined lengths of pipe, rope or band slings should be used to prevent pipe damage. Joined lengths of pipe should never be rolled into trenches, because this may twist the pipe and could cause leaks or pipe damage.

SNAKING OF PIPE

Pipe that is fused or solvent welded and that is less than 6 in. diameter should be snaked to the recommendations that follow. This snaking, or offsetting the pipe with respect to the trench centerline, is necessary in order to provide added pipe length to compensate for any anticipated thermal contraction that will or may take place in the newly joined pipeline.

Snaking is particularly necessary on pipe lengths that may have been joined during the late afternoon heat of a summer day, since their drying time will extend through the cooler temperatures at night when thermal contraction of the pipe could stress the joints to the point that the undried cement will rip. This snaking is also necessary with pipe that is laid in its trench (requiring wider trenches than recommended) and is backfilled with cool earth before the joints are thoroughly dry.

It is not necessary to snake O-ring pipe, since the joint itself allows for the thermal contraction and expansion that may occur.

ANCHORS OR OTHER CONNECTIONS

One rule to remember in designing and installing a plastic underground piping system is that plastic pipe is not designed to be used for any structural applications beyond sustaining normal soil loads and internal pressures up to its hydrostatic pressure rating. Anchors, valve boxes, etc., must be independently supported so that they do not introduce additional bending or shearing stress on the pipe.

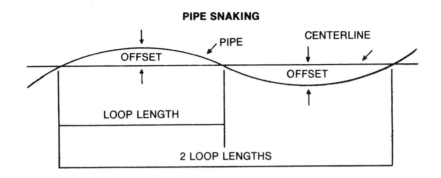

PIPE SNAKING

Loop Offset (in.) for Contraction

Loop Length (ft)	Maximum Temperature Variation, °F (°C), Between Time of Joined Pipe and Final Use									
	10 (5.6)	20 (11.1)	30 (16.7)	40 (22.2)	50 (27.8)	60 (33.4)	70 (38.9)	80 (44.5)	90 (50.0)	100 (55.6)
	Loop offset (in.)									
20	3	4	5	5	6	6	7	7	8	8
50	7	9	11	13	14	16	17	18	19	20
100	13	18	22	26	29	32	35	37	40	42

CONCRETE ANCHORS

If concrete anchors or thrust blocks are going to be poured around plastic pipe, the concrete should be isolated from direct contact with the pipe by means of rubber, etc., wrapped around the pipe prior to pouring. If the purpose for anchoring is to restrain axial movement of the pipe, this can be done by solvent-welding split collars around the outside diameter of the pipe in order to provide a shoulder against the concrete wall. Solvent-welded surfaces between the collar and the pipe's outside diameter must dry for 48 hr before pouring concrete.

CONSTRUCTION OF THRUST BLOCKS

Thrust blocks are anchors placed between pipe or fittings and the trench wall. Blocks can be constructed of available lumber if properly braced. The recommended blocking material is concrete, which is calculated to have a compression strength of 2000 psi. The mixture is one part cement, two parts washed sand, and five parts gravel.

Thrust blocks should be constructed so that their bearing surface or surfaces directly oppose the major force or forces created by the pipe or fitting (see Fig. 5-5). The earth-bearing surface should be undisturbed. While only the simplest of forms is required to restrain the concrete before it sets, care must be taken during its handling so that excessive moisture is not lost to the abutting soil, or that excessive segregation of sand and gravel occurs. These can weaken the support capabilities

of the concrete. Also, the stiffer the concrete mix, the better.

RISERS

Although plastic pipe has excellent weather resistance, it should never be brought aboveground under the following circumstances:

1. If it is expected to provide structural strength, such as supporting an aboveground metal valve.

2. If it is subject to external damage. This could be remedied by sleeving the pipe with an independently and rigidly supported metal pipe.

3. If it is subject to high temperatures, i.e., by nearby steam lines or by summer sun that could lower the pipe's pressure rating below an acceptable level.

PLOWING

"Plowing" is a term used to describe the installation of pipe underground using a vehicle such as a tractor to aid in placing the joined solvent-welded pipe into a trench. If done properly, it is a quick and cost-saving method for installing pipe.

The ability to plow plastic pipe into the ground depends on a number of operating conditions, which can vary widely from job to job. The instructions for plowing are as follows:

1. During hot weather, plastic pipe should be plowed into the ground early in the morning.

2. Absolutely no plowing should be done until the pipe has been given an initial low-pressure test and the joints have dried to the point that they can perform to their 100% hydrostatic pressure rating.

3. Under no circumstances should the pipe be plowed into the ground by "pulling" it in. Instead, it should be plowed in by feeding it over the top of the power vehicle, through the chute on the back of the plow blade, and put into the ground with a laying action. The pipe chute and the blade should be of the same width; wide enough to accommodate the OD of the pipe

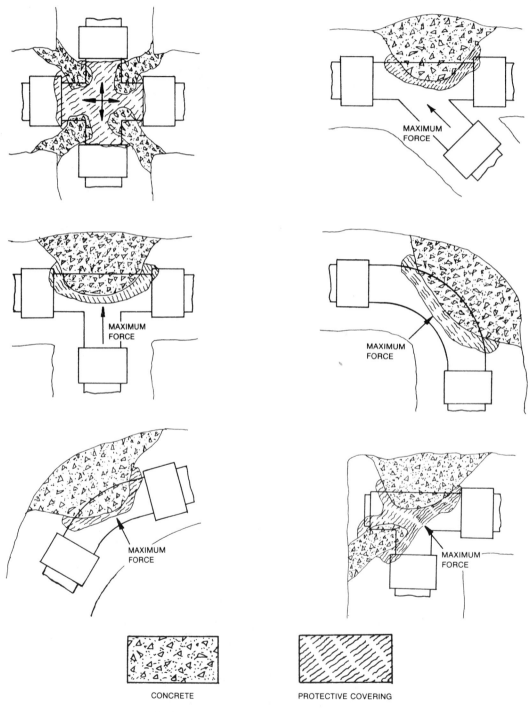

Fig. 5-5. Various types of poured concrete thrust blocking.

joints and to push aside any rocks, stumps, or debris that could damage the pipe. The radius of the chute should be large enough to avoid binding, and excessive stressing and buckling of the pipe. In soil that is quite rocky, it might be advisable to predig the ground with the tractor before laying the plastic pipe.

BACKFILLING COMPACTION

Ideally, during hot weather, backfilling should only be done early in the morning when the line is fully contracted and there is no chance of insufficiently dried joints being subject to contraction stresses.

As already mentioned in the "Trenching" section, in rocky areas there should be a minimum of 4–6 in. of tramped earth or sand beneath the pipe to act as a cushion and for protection of the pipeline from any rocks at the bottom of the trench.

The pipe should first be covered with 6–8 in. of dirt that is free of rocks, debris, or particles larger than 0.5 in. If the subsequent backfill contains rocks, this initial layer should be tamped so that it will act as a cushion for the final backfill. Large sharp rocks that could penetrate through the cushioned layer and damage the pipe should be removed prior to the final backfilling. It is also advisable to maintain a pressure of 15–25 psi in the line during backfilling in order to minimize damage to the pipe.

CONNECTING PIPE SECTIONS

Sections of plastic pipe that have been backfilled or plowed in during the heat of the day should be sufficiently overlapped to allow for contraction and not joined to one another or any stable connection or fitting until the morning following a night of cooling and thermal contraction.

PIPE LOCATING

The location of all plastic pipelines should be accurately and precisely recorded. As an aid for future electronic locating, conductive wire can be trenched or plowed in with the pipe.

ROADWAYS AND RAILROAD TRACKS

When plastic pipe is installed beneath surfaces that are subject to heavy or constant traffic, such as roadways, railroad tracks, etc., it should be run within a metal or concrete casing.

BELOW-GROUND INSTALLATION OF PLASTIC PIPE—POINTS TO REMEMBER

1. Use as narrow a trench as possible to minimize earth loads.

2. Make certain that trench depth is a minimum of 12 in. below the frost line.

3. Snake all small diameter plastic pipe in trenches.

4. Soil surrounding pipe and backfill should be free of any rocks or sharp objects.

5. Be sure the solvent-welded joints are dried before snaking or other pipe movement.

6. If using O-ring pipe, use thrust blocking at any changes in flow direction.

7. Record precisely or use conductive wire to be able to locate accurately underground plastic pipe.

8. Be careful when bringing plastic pipe aboveground. Make sure the pipe aboveground is used properly.

TESTING PLASTIC PIPING SYSTEMS

Whenever possible, plastic piping systems should be tested hydrostatically prior to being put into service. Water is normally used as the testing fluid; do not test with air or gas.

To test a piping system, water is introduced through a 1-in.-diameter or smaller pipe. A cock valve or other means should be available to bleed air from the system. Water is introduced at the lowest point in the system with air being bled off the highest point. A flexible swab or "pig" placed in the line before filling will aid in removing air from the system.

Gradually bring the system up to the desired pressure rating. Test pressures should not exceed the rated operating pressures of the lowest rated component in the system by more than $1\frac{1}{2}$ times.

INITIAL TESTING

An initial low-pressure hydrostatic test should be applied to the joined pipe and fittings before the pipe is backfilled or insulated. This initial pressure test should never exceed 50 psi. The length of time for the low-pressure hydrostatic test should be long enough to determine if there are any minute leaks:

1. If there is complete reliance on the reading of a regular pressure gage, then the pressure may have to be held for several hours in order to detect any minute leaks in a long run of pipe. If the gage indicates that there is a leak, then the pipeline will have to be inspected closely, paying particular attention to the joints, to locate the leak(s) visually.

2. The line can be pressured and without waiting to see if the gage shows a pressure drop, the pipe and joints can be inspected for leaks. It is still recommended that even though no leaks are found visually on the initial inspection, pressure should still be maintained for a reasonable time and the gage rechecked in case a leak is slow in developing.

3. If any leaks have been located and the pipe has been repaired, it is still recommended that the low-pressure test be continued until it is reasonably certain that *all* of the leaks have been located. It is too difficult and costly to locate and repair leaks after installation has been completed.

4. Be careful when testing solvent-cemented joints (PVC, CPVC, ABS), the joints may be tested at 10% of maximum test pressure using Table 5-3 as a guide.

HIGH-PRESSURE TESTING

After completion of the low-pressure test and repair of any leaks, the plastic pipeline may then be high-pressure tested. The high-pressure test should be conducted for at least 12 hr. As a safety precaution while the line is subjected to high-pressure testing, all personnel not working on the test should be kept away from the area in case of pipe rupture.

Table 5-3. Solvent-Cemented Joint Drying Times at 10% of Maximum Test Pressure

| Nominal Pipe Diameter (in.) | Drying Time | | |
	Hot Weather[a] [90–150° F (32–66° C) Surface Temperature]	Mild Weather[a] [50–90° F (10–32° C) Surface Temperature]	Cold Weather[a] [10–50° F (−12–10° C) Surface Temperature]
$^1/_2$–1$^1/_4$	1 hr	1 hr 15 min	1 hr 45 min
1$^1/_2$–2$^1/_2$	1 hr 30 min	1 hr 45 min	3 hr
3–4	2 hr 45 min	3 hr 30 min	6 hr
6–8	3 hr 30 min	4 hr	12 hr

[a]These are atmospheric drying temperatures and should not be confused with atmospheric joining temperature recommendations and limitations.

Table 5-4. Solvent-Cemented Joint Drying Times for 100% Pressure

| Nominal Pipe Diameter (in.) | Drying Time (hr) | | |
	Hot Weather[a] [90–150° F (32–66° C) Surface Temperature]	Mild Weather[a] [50–90° F (10–32° C) Surface Temperature]	Cold Weather[a] [10–50° F (−12–10° C) Surface Temperature]
$^1/_2$–1$^1/_4$	4	5	7
1$^1/_2$–2$^1/_2$	6	8	10
3–4	8	18	24
6–8	12	24	48

[a]These are atmospheric drying temperatures and should not be confused with atmospheric joining temperature recommendations and limitations.

Throughout the testing procedure, do not introduce very high velocities of flow, since excessive water hammer could damage a system.

Care should be taken when testing cemented plastic systems. The system should be fully cured before testing at 100% of desired pressure. In most cases, 24 hr should elapse between completing the last joint and making the test. Table 5-4 gives suggested elapsed times before a system should be tested at full test pressure. Of course, when using threading, flanging, etc., testing can commence immediately.

UNDERGROUND TESTING—POINTS TO REMEMBER

Before testing, all parts of the line should be backfilled and braced sufficiently to prevent movement under pressure.

In setting up a section of line for test, provision for air relief valves should be made.

There are three parts of the line to consider when testing:

1. The run of pipe that must be backfilled sufficiently to prevent movement while under test pressure.

2. Thrust blocks at fittings should be permanent and constructed to withstand test pressure. If concrete thrust blocks are used before testing, enough time must elapse to permit the concrete to set.

3. Test ends should be capped and braced to withstand the appreciable thrusts that are developed under test pressure. (See Table 5-5.)

Where a portion of a line is to be tested and has not yet been tied in to the final source, some other source of water must be found. Table 5-6 shows quantity of water required.

Table 5-5. Thrust at Test Ends for 100 psi Pressure

Pipe Diameter (in.)	Thrust (lb)
1$\frac{1}{2}$	295
2	455
2$\frac{1}{2}$	660
3	985
4	1620
6	3500
8	5930

Table 5-6. Volume of Water Required in Gallons per 100 ft of 160 psi PVC Pipe

Pipe Diameter (in.)	Gal per 100 ft
1$\frac{1}{2}$	13
2	20
2$\frac{1}{2}$	29
3	43
4	70
6	153
8	259

MAKING LEAKAGE TESTS

The purpose of a leaking test is to establish that the section of line to be tested—including all joints, fittings, and other appurtenances—will not leak or that the leakage established is within the limits of the applicable leakage allowance. Leakage, if any, is usually found at joints, saddles, valves, transition fittings, adapters, and other fittings, but rarely in the pipe itself.

For aboveground installations, no leakage is tolerated; however, for certain underground lines handling water, the engineer may establish allowable underground leakage rates and indicate procedures for testing. If not, Table 5-7 should be used to determine how much leakage is allowable in underground water lines.

In setting up a section of line for test, air relief valves should be provided. Air trapped in the line during testing will affect test results and can cause damage to the pipeline.

Table 5-7. Underground Water Lines Leakage Allowance— Gal per 100 Joints per Hour

Pipe Diameter (in.)	Test Pressure (psi)				
	50	100	150	200	225
	Gallons per 100 Joints per hour				
1$\frac{1}{2}$	0.24	0.34	0.41	0.48	0.51
2	0.30	0.42	0.52	0.60	0.63
2$\frac{1}{2}$	0.36	0.51	0.63	0.72	0.77
3	0.44	0.62	0.76	0.88	0.93
4	0.57	0.80	0.98	1.13	1.20
6	0.83	1.18	1.45	1.69	1.77
8	1.08	1.53	1.88	2.17	2.30

Normal operating pressure is usually applied for tests. This should be maintained as constantly as possible throughout the period of test. Measurement of the amount of additional water pumped in during the test provides a measurement of the amount of leakage, if any.

TESTING PLASTIC PIPE—POINTS TO REMEMBER

1. Do not test plastic pipe with air or other gases.

2. Test below-ground piping systems before backfilling to minimize time and money loss in locating and repairing possible leaks.

3. The test pressure should be no more than $1\frac{1}{2}$ times the designed maximum operating pressure or the designed pressure rating of the pipe or any piping component, whichever is lowest.

4. It is imperative to remove air from lines to be tested.

5. Do not impart any large surges of pressure while pumping up the piping system to be tested.

6. While testing at high pressures, it is recommended that a *minimum* amount of personnel be present at the testing site for safety reasons.

7. Do not test with fluid velocities to exceed 5 ft/sec.

UNDERGROUND PIPING HAZARDS

INFILTRATION

Infiltration is a process that allows external matter to enter and contaminate the liquid being transmitted by an underground pipe. Properly solvent-cemented and heat-fused joints of plastic piping permit no infiltration. Mechanically joined pipe tends to be more susceptible to infiltration. Pipe deterioration, broken pipes, leaking gaskets, and improperly installed pipe account for most infiltration.

LEACHING

Leaching can be a chemical or physical process in which pipe wall matter or joining material substances migrate into the liquid inside the pipe.

The substances or leachates may be organic as leached from elastomer gaskets, plastic piping, or metal pipe coatings, or inorganic from metal-piping and joining materials. The *diffusion* process normally causes organic leaching and *corrosion* initiates inorganic leaching of metal-piping components. There are several studies currently being performed by code agencies and independent laboratories to study the effects of organic and inorganic leaching on potable water lines. If a potential harmful effect to humans is found, additional water treatment or specially formulated or treated piping may be required. However, presently there is no evidence that any NSF approved piping (carrying potable water or other foods) has any harmful effect on human beings.

PERMEATION

Diffusion also causes external organic matter to contaminate piping fluids. The organic matter, which migrates from the soil through the piping, is called permeate, and the process is called permeation. Permeation can occur through gaskets and other joining materials as well as directly through the walls of susceptible piping. Organic materials (usually low-molecular-weight organic compounds such as gasoline, oil, bromides, and toluenes) are the culprits in soils; this material may permeate water pipes and cause potential harm to humans. There are many tests that have been and are being performed by piping associations, universities, and code agencies to be able to better predict under what conditions permeation is most likely to occur. Until additional scientific work is completed on pipe permeation, the general rule being used today is not to use *any* potable water piping material in areas of contaminated soil. Localized soil testing will increasingly be performed in both residential and commercial building sites to ensure that no contamination exists.

DOUBLE-CONTAINMENT SYSTEMS

Recently, there has been a vigorous campaign by many consumer and governmental groups to prevent contamination of drinking water reservoirs and aquifers. Double-containment systems of underground piping and tanks for handling toxic and harmful substances are being mandated in several areas of the country. Not only new construction will require a pipe-inside-a-pipe or a tank-inside-a-tank installation, but retrofitting of existing potentially damaging fluid-handling systems will also be required. The design principle of double-contained piping systems is that a pipe transporting the liquid is contained in an outer pipe. If there is any leakage, the leakage will be contained in the outer nonpressurized pipe and will be detected by a leak-sensing device. The sensing mechanism will signal an operator that there is a leak and assist the operator in determining the leak's location. Figures 5-6 and 5-7 show examples of double-containment piping systems.

Fig. 5-6. Double-containment system of polypropylene; pipe, spacers, and couplings are shown. (Courtesy of Asahi/America.)

Fig. 5-7. Double-containment polyester thermosetting piping system. (Courtesy of Smith Fiberglass.)

REPAIRING PLASTIC PIPING SYSTEMS

Repair of plastic piping systems is not difficult and does not require many expensive or sophisticated tools. In fact, plastic pipe leaks, especially in thermoplastic pipe, can be fixed more quickly and successfully than any other material.

THREADED SYSTEMS

Leaks in threaded joints could be caused by improperly tightening joints. If the joint is loose, use a strap wrench to tighten $1\frac{1}{2}$ turns past hand tight. Overtightening a joint will overstress a fitting and can cause cracking. If a fitting is overstressed, a new one should be installed. Use Teflon tape or lubricant on all joints, since this will not only lubricate but aid in sealing. A specially compounded sealant for plastic pipe, Rectorseal®, is used regularly by many contractors.

If a leak occurs when threading dissimilar materials, check the temperature differential in the system; plastics expand at a faster rate than metal. If the system were to increase in temperature by 40° F (22.2° C), the plastic male fitting would tighten into a metal female fitting and thus repair the leak. Conversely, if the temeprature were to decrease by 40° F (22.2° C), the plastic male fitting might loosen in a metal female fitting and cause a leak. Expansion and contraction must be taken into consideration when joining dissimilar materials.

FLANGING SYSTEMS

Leaks at flange faces are usually caused by over and under torquing of bolts. A torque wrench must be used to obtain proper torque. Another reason for leaks could be excessive stress on a flange due to misalignment in a system. Repairing this type of leak requires the revamping of a section of piping to reduce the leak, or the use of long steel bolts to take up the stress, thereby releasing the stress on the plastic flange.

Check the gasketing material if leaks persist; it might not be thick enough. A $\frac{1}{8}$-in.-thick full-face gasket is recommended, preferably of neoprene, Viton, or another proven elastomer.

REPLACING DAMAGED PVC PIPE

For the replacement of damaged underground PVC pipe with a new pipe length, a PVC double-bell coupling is available to simplify and speed-up repair oeprations. The replacement material can consist of a length of pipe with two spigot ends plus two double-bell couplings, or a length with an integral bell plus one double-bell coupling.

In exposing the damaged area, enough of the line should be excavated so that the pipe can be flexed both to aid in handling the damaged area and to insert the replacement material.

Installing a Replacement Section with a Double-Bell Coupling on Each End

1. Cut out the damaged area. Be sure the section cut out includes all damage and leaves a long-enough gap so that a replacement length with a double-bell coupling mounted on each end, as shown in Fig. 5-8, will fit into the gap.

2. Bevel the cut ends of the pipeline using a beveling tool and methods described previously. Put reference marks on these ends.

3. Determine the necessary length of the replacement pipe as shown in Fig. 5-8, then cut it to the proper length. In determining the proper length of replacement pipe to be cut, measure the gap dimension, multiply it by 2, and subtract the result from the length of the cut out section.

After cutting and beveling the ends of the replacement pipe, mount both couplings on the ends so that they are in the position as shown in Fig. 5-8, at top.

4. Insert the replacement assembly in the line and slide the couplings into proper position so that each coupling is centered over the gap and midway between both sets of reference marks, as shown in Fig. 5-9.

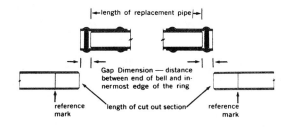

Fig. 5-8. Measuring a replacement section of pipe. (Courtesy of Johns-Manville Corp.)

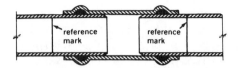

Fig. 5-9. Inserting a pipeline using two double-bell couplings. (Courtesy of Johns-Manville Corp.)

Procedure when a Length in Integral Bell End Plus One Double-Bell Coupling Is Used

Accomplish necessary cutting and beveling as described previously, taking care to allow for the gap dimension on one end only in measuring the pipe length. Locate a reference mark on one cut end, as previously described. Mount the coupling on the cut end of the replacement length so that only the gap dimension overhangs. Insert in the pipeline as follows:

1. Complete the integral-bell joint first by pushing the bell end onto the cut spigot end. Use the reference mark to position bell end properly.

2. Slide the double-bell coupling onto the cut end of the line and center it over the gap.

THERMOPLASTICS (CEMENTING/FUSION)

Leaks in cemented or fusion-type piping systems are a result of poor installation techniques. However, because of the ability of thermoplastics to be welded, *small* leaks may be repaired easily and successfully. If the leak is a steady stream, replace the pipe and fitting.

Most leaks occur at joints. If a leak should occur at any other place in the system, cut the section or fitting out and replace it.

Before making repairs in a plastic piping system, that section of the system to be worked on should be drained and allowed to dry completely. The repair is then made by an experienced plastic welder who needs to be supplied with an electric source and, possibly, a compressed air source if there is no integral air source in the type of welding gun used. To train workers to weld plastics is not difficult and should take no more than 1 hr of training by an experienced welder. In an accessible location, to repair a 3-in. joint, three beads of weld should take no more than 20–30 min. Table 5-8 summarizes welding characteristics of five commonly used plastic piping materials.

WELDING PROCEDURE

1. Welding surfaces must be free of dirt, oil, moisture, and loose particles of plastic material. When backwelding solvent-welded joints, allow cement to cure for 6 hr before welding. As seen in Fig. 5-10, remove all excess cement residue with a knife, emery cloth, or wire brush before welding. Do not backweld a joint while it is leaking, since the moisture will prevent a good bond. THE PIPELINE SHOULD BE DRAINED PRIOR TO BACKWELDING.

2. Cut the plastic welding rod (which must be of the identical material as the pipe) at a 45° angle with a pair of snippers. (See Fig. 5-11.)

3. Welding thermoplastics requires hot air or gas at temperatures recommended by the manufacturer. (See Table 5-8.) Maintain uniform heat and pressure on the rod while welding. (See Fig. 5-12.) Too much heat will char, melt, or distort the material; too much pressure on the rod tends to stretch the weld bead, which could result in cracks in the weld after it cools. When welding PVC, CPVC, or polypropylene, the rod should be held at an angle of 75–90° to the joint. When welding polypropylene, the rod will bend in an arch when proper pressure is applied.

4. Welds should never be spliced by overlapping side by side. When terminating a weld, the bead should be lapped on top of (never alongside) itself for a distance of $3/8–1/2$ in. as in Fig. 5-13. A single-drip leak usually can be repaired

Table 5-8. Welding Requirements and Properties of Five Commonly Used Plastic Piping Materials

Requirements and Properties	Plastic Materials				
	PVC Type I	CPVC	PP	PVDF	PE
Welding temperature	525° F (274° C)	525° F (274° C)	575° F (302° C)	585° F (307° C)	550° F (288° C)
Welding gas	Air	Air	Nitrogen	Air	Nitrogen
Odor under flame	Acid	Acid	Wax	Acid	Wax
Position of rod	90°	90°	75°	75°	75°
Remarks	Low weld strengths	Low weld strengths; difficult to weld	May splash; reduce airflow; fast welding		Fast welding

Fig. 5-10. Removing excess cement residue before making a weld.

Fig. 5-11. Cutting plastic welding rod.

Fig. 5-12. Welding the rod.

Fig. 5-13. Second bead of weld being applied.

with a single-bead weld; more serious leaks require full-fillet welds (Fig. 5-14), usually three beads, but up to five beads in large-diameter pipe. When making multiple-layer welds, allow ample time for each pass to cool before proceeding with final welds.

Other methods besides welding have been used successfully to stop leaks, such as repair couplings, repair saddles, and flanged spool pieces. Generally, these methods are used in damaged pipe where there are no fittings or changes in direction.

A unique repair system has been used with some success in solvent-cemented joints. The piping system is emptied of liquids and a vacuum pump is used to create a vacuum in the system. The leaky joints have cement squeezed into any noticeable voids by using a squeezable plastic ketchup dispenser. The vacuum sucks the cement into the voids and may seal the leaks. Backwelding, however, is recommended in addition to vacuum sealing.

THERMOSETS

Thermoset pipe cannot be welded. If a leak should occur at a joint in a fiberglass-reinforced plastic system, in almost every case the section should be replaced, especially in *medium- to high-pressure-rated systems.* Joint areas have been successfully overwrapped by a "Patch-Kit" method, but at joint areas it is difficult to get a homogeneous and fully adhered bond. If the manufacturer has experience in overwrapping joints and recommends this procedure for repairing leaks, make certain that a manufacturer's representative supervises the repairs.

Straight runs of pipe with leaks can be repaired by cutting out the damaged section and installing a cemented repair coupling or flanged-spool piece.

Fig. 5-14. Repairing a serious leak with a full-fillet weld. (Courtesy Chemtrol Div. of Nibco.)

VALVES

The repairing of plastic valves is no different than with any other valves. O-rings, packing, diaphragms, gaskets, and other wearing parts may be replaced on most plastic valves. Some valves are inexpensive, and the decision whether to repair or discard the valve depends more on the labor and parts cost, versus the cost of a new valve.

All plastic valve manufacturers carry a sufficient parts inventory to deliver valve parts quickly as required. It is recommended that wearing parts such as O-rings or diaphragms be kept in inventory by the valve user in case of needed repair or a planned maintenance program.

REPAIRING PLASTICS—POINTS TO REMEMBER

1. To minimize repairs, do not take short cuts on installing plastic piping systems.

2. Do not overtighten threaded joints. Either replace threaded joints or use Teflon sealer.

3. Do not overtorque bolts on plastic flanges. Either check stress on system or use new and thicker gasket.

4. To repair solvent-cemented joints, the piping systems must be dry, especially the area to be welded.

5. Follow welding instructions completely to ensure a lasting repair of solvent-cemented joints.

6. Use O-ring repair couplings when repairing underground O-ring PVC pipe.

7. When repairing thermosets, use "patch kits" cautiously and, if possible, have a manufacturer's representative present to supervise the repair.

8. Carry spare wearing-valve parts in stock, especially in a critical application.

6
Cost Comparisons

INTRODUCTION

The goal of a specifying engineer is to obtain maximum performance level at minimum cost. The engineer's decision is based on four major considerations:

1. Material performance.

2. Material cost.

3. Initial installed labor cost.

4. Long-term maintenance cost.

Usually, the initial material selection phase discovers several piping materials that will perform adequately, from a technical standpoint. Material selection has been covered in previous chapters, but the ranking of these several acceptable materials by their costs must now be considered.

To forecast a true economic analysis of a proposed piping system is difficult; however, by assuming particular parameters, a useful analysis may be done.

MATERIAL COST

There are occasional material shortages and we do live in an inflationary environment; hence, prices of materials fluctuate greatly. In fact, most suppliers guarantee prices for no more than 6 months. All cost indices given in this chapter's tables are by price comparisons as of July 1987. Contractor pricing is used in developing each index; keep in mind that discounts may vary widely from supplier to supplier and often depend on the quantity of pipe, fittings, and valves on a particular job.

For aboveground piping, PVC-80 is indexed @ 1.00; underground piping is based on polyethylene having an index of 1.00. These indices are used in order to correlate materials of the most commonly used plastic piping material. Several sizes of pipe, fittings, and valves are tabulated, illustrating existing pricing trends.

LABOR COSTS

To include material cost alone in evaluating piping systems is nonobjective and illogical in today's world where the cost of labor is so high—50–70% of the total cost of a complex piping system. The cost of labor, therefore, must be included in any evaluation for installing a piping system.

The labor cost of installing a piping system depends on the following factors: (a) type of material; (b) selection of joining method; (c) number of joints; (d) piping environment; (e) labor rate; and (f) labor productivity. Articles from B. F. Goodrich, Dow, NIBCO, Delta Faucet, Tuthill's *Installed Costs of Corrosion-Resistant Piping, Chemical Engineering* Part 1, March, 1986, Corban Industries, and field experience have been used in formulating labor and material costs.

TYPE OF MATERIAL

Not only are there many materials available for use in piping fluids, but for each particular material, there are various sizes and wall thicknesses to choose from. For example, certain sizes of PVC pipe are readily available in seven or eight different wall thicknesses or pressure ratings. However, only the more commonly used piping materials, pressure ratings, and sizes have been used in the evaluations. Tables 6-1 and 6-2 describe the materials used for cost comparisons for different piping. Two of the cost savings that are not shown in the cost comparisons are freight savings and the savings in labor to unload a truck of plastic piping versus nonplastic pipe. In many cases thousands of dollars may actually be saved by taking advantage of this unique property of plastic piping—its light weight. The need for less rigging equipment to install plastic piping has been accounted for in the cost indices.

SELECTION OF JOINING METHOD

For any given material, there may be several joining methods available, each of which could directly affect the labor cost for any particular piping system. But for our purposes, the more commonly used joints for a particular application and material have been selected for cost comparisons. Tables 6-1 and 6-2 list the type of joints selected for cost comparisons.

Table 6-1. Description of Aboveground Pressure Piping Materials

Material	Schedule or Pressure Rating	Pipe Length (ft)	Pipe Ends	Type of Joining
Alloy G	5S	20	Butt weld	Welding
Aluminum	40	20	Butt weld	Welding
Black iron	40	20	Threaded	Threading
Carbon steel	40	20	Threaded	Threading
Copper	Type L	20	Plain	Solder/braze
CPVC	80	20	Plain	Cemented
Galvanized steel	40	20	Threaded	Threading
PVC	80	20	Plain	Cemented
Polypropylene	80	20	Plain	Fusion
PVDF (Kynar®)	80	20	Plain	Fusion
Polypropylene lined	150 psi	10	Flanged	Flanging
PVDF lined	150 psi	10	Flanged	Flanging
PTFE (Teflon) lined	150 psi	10	Flanged	Flanging
RTRP polyester	150 psi	20	Plain	Cemented
Red brass (85)	125 psi	12	Threaded	Threading
Stainless steel-304	5S	20	Butt weld	Welding
Stainless steel-316	5S	20	Butt weld	Welding
Polyethylene	160 psi	20	Plain	Fusion
20 Cr–25 Ni–6 Mo	5S	20	Butt weld	Welding
Rubber (natural) lined	150 psi	10	Flanged	Flanging

NUMBER OF JOINTS

Normal stocked lengths of pipe have been used to determine the number of joints in our cost comparison. The aboveground piping system is a simple piping arrangement where there is 400 ft of pipe with six 90° ells and six tees. There are 30 joints using 20-ft sections of pipe and 40 joints using 10-ft sections of pipe. For the underground piping system, the number of joints correspond to the pipe lengths of a straight run of 500 ft of pipe. For example, there would be 50 joints for 10-ft lengths of pipe. There are no fittings included in the underground piping system. Joints can be prefabricated and then installed in the field, which many times is the preferred practice from a cost and dependability standpoint. Many industrial piping systems are available completely flanged for ease in field assembly. However, cost comparisons used in our tables assume all joints to be field made and installed unless a piping system is manufactured only with flanged ends, such as lined piping.

PIPING ENVIRONMENT

All of the tables assume handling and erection of pipe within 200 ft of the stockpile at grade level with up to a maximum installation height of 10 ft and a maximum trench depth of 8 ft. It is assumed that space exists at the site with easy access for piping installation.

LABOR RATES AND PRODUCTIVITY

Labor rates vary with location and time of year. For our analysis we used a labor rate figure of $40.00 per man hour, average labor cost. This includes the contractor's overhead, insurance, and taxes. Labor costs should also be adjusted to reflect local conditions.

All estimates are based on an 8-hr day with an 80% efficiency factor used to determine productive work time. Productivity varies drastically with a work crew's experience, type of supervision, etc. Hours and efficiency must be adjusted to reflect local conditions.

Table 6-2. Description of Below Ground Piping Materials

Materials	Schedule of Pressure Rating	Pipe Lengths (ft)	Pipe Ends	Type of Joining
Cast iron	Soil—A-74	10	Hub × spigot	Gasket
Cast iron ductile	Class 50	18' 2"	Bell × spigot	Gasket
Clay-vitrified	Extra strength	4"/8"/36" = 5' 12" = 6' 18" = 7'	Bell × spigot	Gasket
Polybutylene	SDR 34.3	20[a]	Plain end	Fusion
Polyethylene[b]	SDR 32.5	20[a]	Plain end	Fusion
Polyethylene[b]	Class 63	20[a]	Bell × spigot	Gasket
Lined concrete	Class 150	4"–24" pipe diam. = 32' lengths 36" pipe diam. = 24' lengths 48"–72" pipe diam. = 16' lengths	Bell × spigot	Gasket
PVC	SDR 35	20[a]	Bell × spigot	Gasket
PVC	SDR 26	20[a]	Bell × spigot	Gasket
PVC	Sch 40	20[a]	Plain end	Cemented
PVC (C-900)	DR 25/Class 100	20	Bell × spigot	Gasket
Reinforced concrete	Class 4	4" = 3' 18"–48" = 8' 8" = 4' 72" = 7½' 12" = 6'	Tongue and groove (cemented to 12 in.)	Push-on
RTRP polyester[c]	50 psi	20	Bell × spigot	Gasket

[a]Most plastic piping may be manufactured and shipped in 40- and 60-ft lengths. For our cost comparisons, 20 ft lengths were used.
[b]Polyethylene pipe is butt-fused up to and including 36 in. pipe; 48 and 72 in. pipe are gasket connections.
[c]RTRP is adhesive cemented up to and including 12 in. and gasketed joined above 12 in.

OTHER ASSUMPTIONS FOR TABLES

How, except for comprehensive and in-depth motion studies, does one determine the labor savings in handling plastic versus nonplastic pipe, or the additional support spacing required for plastic versus nonplastic, or the additional drying time required for solvent joints versus other joining methods? To obtain costs of labor we assume the additional labor savings for handling plastic are offset by additional support spacing requirements. For solvent cementing, the actual joint time has been doubled to correct for drying time required to move the pipe after joining. Table 6-3 determines the man hours for handling various joining methods of pipe sizes from 1 to 12 in.; Table 6-4 lists the man-hours for below ground

joining methods. In estimating the joint times, these additional assumptions have been made:

1. Power tools are to be used where applicable.

2. Joint times include cutting, beveling, and threading (if required).

3. Hangers and labor for hanging pipe are not included.

4. Rentals of trenching equipment are not included.

HOW TO USE THE TABLES

Be aware that costs both for above- and below ground piping systems will vary depending on each application. It is quite possible, for example,

Table 6-3. Estimated Man-Hours for Joining Methods of Aboveground Piping

Type of Joint	1	2	4	6	8	10	12
	Estimated Man-Hours per Joint (hr)						
Welded	0.15	0.30	0.80	1.32	1.72	2.16	2.81
Threaded	0.08	0.18	0.80	2.00	...	...	...
Flanged[a]	0.06	0.11	0.15	0.40	0.50	0.80	1.00
Cemented	0.07	0.15	0.25	0.66	0.86	1.08	1.41
Fusion	0.05	0.10	0.18	0.50	0.70	0.92	1.20
Mechanical	0.04	0.08	0.12	...	...	...	...
Solder/brazed	0.07	0.15	0.50	1.50	2.15	...	...

Pipe Size (in.)

[a]Solid flanges to 2 in. and Van Stone flanges above 2 in.; lined piping comes with solid one-piece flanges.
[b]Compression-type joints (not used in our analysis).

Table 6-4. Estimated Man-Hours for Joining Methods of Below Ground Piping

Type of Joint	4	8	12	18	24	36	48	72
	Estimated Man-Hours per Joint (hr)							
Gasketed	0.12	0.24	0.40	0.55	0.70	0.90	1.20	1.60
Cemented	0.25	0.86	1.41	2.00	2.50	N/A[a]	N/A	N/A
Butt fusion	0.20	0.70	1.20	1.60	2.00	2.75	3.65	4.75
Welded	0.80	1.72	2.81	3.30	4.20	5.40	7.20	9.60
Butt and wrap[b]	Similar to welded joint							

Pipe Size (in.)

[a]N/A = not available.
[b]Used in RTRP systems but not a considered joining method for our comparisons.

that in determining underground piping labor and material costs, geography and time of the year could directly affect trenching, bedding, and even piping costs. Local plumbing codes also play a significant role in determining material applicability in various parts of the country. In essence, do not use the following tables blindly and unquestioningly. The tables are intended only as a *guide* in comparing material, labor, and total installed costs for selected piping products. Probably the surest way to determine actual compar-

ative costs for any installation is to get bids from reliable local contractors who are experienced in installing plastic and nonplastic piping products.

There are many variables that *must* be considered in conjunction with the cost comparison tables (Tables 6-5 and 6-6). For example, in Table 6-5 the installed cost index of 4-in. polyethylene is shown as 0.75 and is less than 4-in. PVC-80 having a 1.00 index. However, if at the job site no economical source of gas or electric power exists for fusion of the polyethylene or the work crew

Table 6-5. Material/Labor Installed Cost Indexes of Commonly

Material	1-in. Pipe			2-in. Pipe			4-in. Pipe		
	Material Index	Labor Index	Installed Index	Material Index	Labor Index	Installed Index	Material Index	Labor Index	Installed Index
Alloy G	NCA[b]	NCA	NCA	27.65	2.62	4.51	18.42	2.14	3.73
Aluminum	4.16	2.15	2.23	3.24	2.46	2.63	3.85	2.00	2.38
Black iron	NCA	NCA	NCA	1.33	1.38	1.38	0.81	1.17	1.13
Carbon steel	1.92	1.45	1.58	1.46	1.38	1.39	1.37	1.17	1.19
Copper	2.11	1.00	1.04	2.42	1.00	1.11	2.91	2.00	2.07
CPVC	2.53	1.00	1.06	2.63	1.00	1.17	2.79	1.00	1.27
20Cr–25Ni–6 Mo	NCA	NCA	NCA	10.51	2.62	3.22	10.65	2.47	3.27
Galvanized steel	1.82	1.45	1.46	1.67	1.40	1.48	1.96	1.49	1.54
Polyethylene	1.12	0.71	0.73	0.94	0.67	0.69	0.98	0.72	0.75
Polypropylene	2.80	0.71	0.79	2.21	0.67	0.79	2.98	0.72	0.94
Polypropylene lined	37.44	3.50	4.80	20.99	3.23	4.57	13.99	2.21	3.36
PVC-40	0.79	1.00	0.99	0.75	1.00	0.98	0.62	1.00	0.96
PVC-80	1.00	1.00	1.00	1.00	1.00	1.00	1.00	1.00	1.00
PVDF[a]	18.66	0.71	1.40	16.35	0.67	1.86	19.69	0.72	2.57
PVDF lined	61.91	3.81	6.05	32.17	3.23	5.42	23.33	2.21	4.27
PTFE lined	80.60	4.73	7.65	46.20	3.23	6.48	34.13	2.21	5.34
Red brass	8.89	1.45	1.74	8.75	1.38	1.94	10.37	1.17	2.07
RTRP	15.89	1.00	1.57	7.12	1.00	1.50	3.87	1.00	1.28
Rubber lined	NCA	NCA	NCA	NCA	NCA	NCA	19.20	3.53	5.06
Stainless steel 304	6.53	2.15	2.32	4.66	2.46	2.63	3.67	2.00	2.16
Stainless steel 316	8.92	2.15	2.41	6.38	2.46	2.76	4.84	2.00	2.28

[a]PVDF = Sch 80 pipe and fittings from 1 to 6 in.; 8 to 12 in. are 150-psi-rated piping system.

[b] = not commonly available.

Table 6-6. Material/Labor Installed Cost Indexes of

Material	4-in. Pipe			8-in. Pipe			12-in. Pipe			18-in. Pipe		
	Material Index	Labor Index	Installed Index	Material Index	Labor Index	Installed Index	Material Index	Labor Index	Installed Index	Material Index	Labor Index	Installed Index
Cast iron	3.75	1.20	2.61	3.15	0.69	2.03	3.29	0.67	2.23	2.32	0.69	1.83
Cast iron, ductile	7.58	0.67	4.51	3.50	0.39	2.08	2.63	0.37	1.72	2.27	0.39	1.70
Clay	2.83	2.40	2.64	1.65	1.37	1.52	1.62	1.12	1.42	2.14	0.99	1.79
Concrete lined	NCA[a]	NCA	NCA	NCA	NCA	NCA	NCA	NCA	NCA	2.12	0.22	1.55
Concrete reinforced	2.18	4.00	2.99	0.98	1.71	1.31	1.00	1.12	1.05	0.65	0.87	0.72
Polybutylene	2.43	1.00	1.79	2.54	1.00	1.84	2.44	1.00	1.86	1.79	1.00	1.55
Polyethylene	1.00	1.00	1.00	1.00	1.00	1.00	1.00	1.00	1.00	1.00	1.00	1.00
PVC-SDR 35	1.00	1.25	1.11	1.08	1.23	1.15	1.20	1.18	1.19	1.32	1.25	1.30
PVC-SDR 26	1.03	1.25	1.13	1.16	1.23	1.19	1.34	1.18	1.28	1.81	1.25	1.64
PVC-Sch 40	1.93	1.25	1.63	1.55	1.23	1.40	1.32	1.18	1.26	NCA	NCA	NCA
PVC-C900	1.60	1.25	1.44	1.55	1.23	1.40	1.47	1.18	1.35	NCA	NCA	NCA
RTRP	6.37	1.25	4.09	6.17	1.23	4.67	5.80	1.18	3.94	4.11	0.35	3.25

[a]NCA = not commonly available.

is much more familiar with PVC installation, the proper decision might be to use PVC-80 piping. Price changes will affect the tables as will any special discounts a buyer may obtain for a particular material. In the case of labor, local conditions must be known and the tables adjusted to show any change from the stated assumptions. *The tables do show relative cost comparison between materials* and if all assumed conditions hold, the tables will show how different materials vary with each other from an economic analysis.

The most pertinent data in the tables are the installed cost indices, which combine material and labor costs—the most meaningful criteria in comparing pipe materials.

VALVES

Valving costs are much more difficult to compare than pipe and fitting costs. The almost un-

Used Aboveground Pressure Piping Materials (PVC 80 = 1.00)

6-in. Pipe			8-in. Pipe			10-in. Pipe			12-in. Pipe		
Material Index	Labor Index	Installed Index	Material Index	Labor Index	Installed Index	Material Index	Labor Index	Installed Index	Material Index	Labor Index	Installed Index
11.90	2.41	3.77	12.57	2.90	4.55	17.27	3.47	7.05	16.37	2.48	5.93
NCA	NCA	NCA	NCA	NCA	NCA	NCA	NCA	NCA	NCA	NCA	NCA
0.75	1.16	1.10	0.78	1.20	1.13	0.57	1.01	0.90	0.66	0.88	0.83
1.19	1.16	1.17	1.15	1.20	1.19	0.70	1.01	0.93	0.70	0.88	0.84
4.92	2.27	2.65	7.09	2.50	3.28	NCA	NCA	NCA	NCA	NCA	NCA
2.95	1.00	1.42	3.70	1.00	1.63	3.85	1.00	1.74	NCA	NCA	NCA
5.63	2.35	2.82	5.76	2.82	3.32	8.01	2.70	4.08	7.55	2.44	3.71
1.81	1.77	1.78	NCA	NCA	NCA	NCA	NCA	NCA	NCA	NCA	NCA
1.12	0.76	0.81	1.42	0.81	0.91	0.85	0.85	0.85	0.96	0.85	0.88
3.17	0.76	1.11	1.43	0.81	0.92	1.00	0.85	0.89	1.37	0.85	0.98
11.75	2.70	4.00	11.87	2.56	4.15	7.69	2.57	3.82	7.98	2.50	4.48
0.54	1.00	0.93	0.58	0.85	0.80	0.81	0.80	0.80	0.68	0.85	0.81
1.00	1.00	1.00	1.00	1.00	1.00	1.00	1.00	1.00	1.00	1.00	1.00
23.08	0.76	3.96	5.48	0.81	1.61	3.51	0.85	1.54	4.37	0.85	1.73
22.30	2.70	5.52	20.81	2.56	5.67	20.46	2.57	7.21	19.16	2.50	6.64
32.46	2.70	6.97	33.67	2.56	7.86	20.13	2.57	7.13	21.79	2.50	7.30
NCA	NCA	NCA	NCA	NCA	NCA	NCA	NCA	NCA	NCA	NCA	NCA
3.41	1.00	1.35	3.39	1.00	1.41	2.05	1.00	1.27	1.98	1.00	1.24
15.46	3.95	5.60	10.16	3.97	4.99	7.73	3.81	4.83	3.41	3.45	3.44
3.56	2.27	2.46	3.28	2.74	2.83	2.96	2.53	2.64	2.50	2.35	2.39
4.74	2.27	2.62	4.20	2.74	2.99	3.64	2.53	2.82	3.24	2.35	2.56

Commonly Used Below Ground Piping Materials (Polyethylene = 1.00)

24-in. Pipe			36-in. Pipe			48-in. Pipe			72-in. Pipe		
Material Index	Labor Index	Installed Index	Material Index	Labor Index	Installed Index	Material Index	Labor Index	Installed Index	Material Index	Labor Index	Installed Index
1.87	0.70	1.59	1.58	0.65	1.43	1.88	2.00	1.90	NCA	NCA	NCA
1.92	0.39	1.55	1.54	0.37	1.35	1.88	1.11	1.77	NCA	NCA	NCA
2.22	1.00	1.94	NCA	NCA	NCA	NCA	NCA	NCA	NCA	NCA	NCA
1.88	0.22	1.48	1.28	0.27	1.09	1.04	1.25	1.07	0.76	1.28	0.81
0.56	0.88	0.64	0.51	0.82	0.56	0.56	2.52	0.84	0.62	2.68	0.81
2.46	1.00	2.11	NCA	NCA	NCA	NCA	NCA	NCA	NCA	NCA	NCA
1.00	1.00	1.00	1.00	1.00	1.00	1.00	1.00	1.00	1.00	1.00	1.00
1.37	1.25	1.34	NCA	NCA	NCA	NCA	NCA	NCA	NCA	NCA	NCA
1.90	1.25	1.74	NCA	NCA	NCA	NCA	NCA	NCA	NCA	NCA	NCA
NCA	NCA	NCA	NCA	NCA	NCA	NCA	NCA	NCA	NCA	NCA	NCA
NCA	NCA	NCA	NCA	NCA	NCA	NCA	NCA	NCA	NCA	NCA	NCA
3.24	0.35	2.76	2.22	0.33	1.91	2.14	1.00	1.98	1.70	1.00	1.64

limited trim materials, variety of sealing, and seating synthetics as well as the differences in pressure ratings and type of end connections, make a meaningful comparison of plastics versus metals an arduous task. However, Table 6-7 offers a cost comparison by materials cost only. The valve pricing is based on selecting the least costly shutoff valve per size per material. For example, if a 4-in. butterfly valve was less costly than a 4-in. ball valve, the butterfly valve price was used for our cost index. No consideration was given to backflow prevention or throttling valves unless the throttling valve was less costly than an on/off valve. The price index also used the least costly available sealing elastomers and the preferred pipe joining connection. All valves above 3 in. are flanged connections.

Although Table 6-7 shows cost comparisons from a material standpoint, it does not determine the total installed cost comparison of plastic and metal valves. The buyer and installer must consider the fact that plastic valves are lighter in weight than metal valves and, in sizes above 2 in. in particular, would require much less heavy duty support; also, no rigging equipment would be

Table 6-7. Material Cost Indexes of Commonly Used Industrial 150-psi-Rated Valves (PVC = 1.00)

Material	Size (in.)						
	1	2	4	6	8	10	12
Bronze	0.54	0.64	0.64	0.88	0.96	0.76	0.77
Carbon steel	0.77	1.00	3.65	6.38	5.72	6.16	8.47
Cast iron	NCA[a]	1.85	0.61	0.88	0.96	0.76	0.77
CPVC	1.38	1.50	1.64	NCA	NCA	NCA	NCA
Forged steel	3.69	4.11	3.49	6.98	8.70	12.06	11.20
Monel	24.46	24.50	18.78	34.01	NCA	NCA	NCA
Polypropylene	1.31	1.50	1.64	5.18	5.85	6.60	NCA
PVC	1.00	1.00	1.00	1.00	1.00	1.00	1.00
PVDF	4.31	4.79	3.55	5.84	7.93	6.60	5.85
Stainless steel 304	1.62	1.86	4.76	8.83	12.83	15.23	16.51
Stainless steel 316	1.69	2.00	5.22	9.45	8.93	9.50	10.89
Glass-reinforced vinylester	13.38	10.71	2.63	4.09	3.27	3.18	2.76

[a]NCA = not commonly available.

needed to install plastic valving. There are also savings in not having to coat, paint, or treat the exterior of the plastic valve to prevent environmental attack. As with all plastic fluid-handling products, the flow characteristics of plastics have less friction loss compared to most nonplastic materials, allowing significant power savings.

7
Applications

INTRODUCTION

Plastic piping applications date from the mid-1930s when PVC piping was first utilized, as it still is, for sanitary drainage in Germany. Plastic piping materials are now used in almost every industry for all conceivable applications.

By far, the most commonly used plastic piping applications are nonindustrial such as irrigation piping, drain–waste–vent piping for homes, water mains and service lines, natural gas transmission service lines, and home fire and lawn sprinkler systems. In these markets, plastic piping is displacing metals as the preferred material.

Several industrial applications of plastics will be presented in this chapter, but these are merely a sampling since there is no limit to the possible new applications of plastics in industry. For more detailed listings of plastic pipe, valves, and fittings applications, contact plastic piping system manufacturers, resin producers, and plastic piping distributors, many of whom were kind enough to allow the use of their installation photographs.

CHEMICAL PROCESSING

There is no major segment of the chemical processing industry that is not using plastic piping systems (see Table 7-1). Because of the many advantages of plastics listed in Chapter 1, the U.S. usage of plastic piping should increase tenfold or more in the next few years. However, owing to lack of education and training, many design and plant engineers are ignorant of the features of plastic piping and hence use less efficient and more costly piping materials. One of the objectives of this text and other similar publications has been to better educate industry in the proper design and usage of plastic piping systems. (For some typical applications, see Figs. 7-1–7-4.)

FOOD PROCESSING

Most plastic piping materials are approved by the National Sanitary Foundation (NSF), and receive Food and Drug Administration (FDA) approval when required. The purity of the end-product in any food processing application is critical. Plastics fit the bill beautifully. The inertness and smooth flow rates of the plastic piping material make it a preferred material in many food processing applications. Some of these applications (see Fig. 7-5) are listed as follows:

   I. Mustard, Vinegar, and Pickle Plants
     A. Brine line
     B. Vinegar lines
     C. Mustard process lines
     D. Auxiliary water and air lines
  II. Rice Mills
     A. Chemical bleaching lines
 III. Citrus Industry
     A. Citrus acid lines
     B. Low-temperature process lines
 IV. Corn Processing
     A. Corn syrup lines
     B. Hot corn liquors
  V. Salt Manufacturing Plants
     A. All low-temperature process lines

Table 7-1. Markets Defined of the Chemical Process Industries

Chemicals (Including Petrochemicals)
Alkalis and chlorine
Industrial gases
Industrial inorganic chemicals
Synthetic rubber (vulcanizable, elastomers)
Gum and wood chemicals
Cyclic, crudes, and intermediates
Industrial organic chemicals
Chemicals and chemical preparations

Drugs and Cosmetics
Biological products
Medicinal chemicals and botanicals
Pharmaceutical preparations
Perfumes, cosmetics, and other toilet
preparations

Explosives and Ammunition
Explosives
Small arms ammunition
Ammunition, excluding small arms
Ordnance and accessories

Fats and Oils
Cottonseed oil mills
Soybean oil mills
Vegetable oil mills
Animal and marine fats and oils
Shortening and cooking oils

Fertilizers and Agricultural Chemicals
Nitrogenous fertilizers
Phosphatic fertilizers
Fertilizers, mixing only
Agricultural chemicals

Foods and Beverages
Condensed and evaporated milk
Wet corn milling
Cane sugar, except refining only
Cane sugar refining
Beet sugar
Malt liquors
Malt
Wines, brandy, and brandy spirits
Distilled, rectified, and blended liquors
Flavoring extracts and syrups
Roasted coffee
Food preparations

Leather Tanning and Finishing
Lime and Cement
Cement hydraulic
Lime

Man-made Fibers
Cellulosic man-made fibers
Synthetic organic fibers, except cellulosic

Metallurgical and Metal Products
Electrometallurgical products
Primary smelting and refining of copper
Primary smelting and refining of lead
Primary smelting and refining of zinc
Primary production of aluminum
Primary smelting and refining of nonferrous
metals
Secondary smelting, refining, and alloying
of nonferrous metals and alloys
Enameled iron and metal sanitary ware
Electroplating, plating, polishing, anodizing,
and coloring
Coating, engraving, and allied services

**Paints, Varnishes, Pigments, and Allied
Products**
Inorganic pigments
Paints, varishes, lacquers, and enamels

Petroleum Refining and Coal Products
Petroleum refining
Paving mixtures and blocks
Asphalt felts and coatings
Lubricating oils and greases
Coke and by-products

**Plastics Materials Synthetic Resins, and
Nonvulcanizable Elastomers
Rubber Products**
Tires and inner tubes
Rubber footwear
Reclaimed rubber
Fabricated rubber products

**Soap, Glycerin, Cleaning, Polishing, and
Related Products**
Soap and other detergents, except specialty
cleaners
Specialty cleaning, polishing, and sanitation
preparations, except soap and detergents
Surface active agents, finishing agents,
sulfonated oils, and assistants

Stone, Clay, Glass, and Ceramics
Flat glass
Glass containers
Pressed and blown glass and glassware
Brick and structural clay tile
Ceramic wall and floor tile
Clay refractories
Structural clay products
Vitreous china plumbing fixtures, china,
and earthenware fittings, and bathroom
accessories
Vitreous china table and kitchen articles
Fine earthenware (Whiteware) table and
kitchen articles
Porcelain electrical supplies
Pottery products
Gypsum products
Abrasive products
Asbestos products
Steam and other packing and pipe and
boiler covering
Minerals and earth, ground or otherwise
treated
Mineral wool
Nonclay refractories
Nonmetallic mineral products

Wood, Pulp, Paper, and Board
Wood preserving
Pulp mills
Paper mills, except building paper mills
Paperboard mills
Paper Coating and glazing
Building paper and building board mills

Other Chemically Processed Products
Part of broad woven fabric mills, wool:
dyeing and finishing only
Finishers of broad woven fabrics of cotton
Finishers of broad woven fabrics of man-
made fiber and silk
Dyeing and finishing textiles
Artificial leather, oilcloth, other impregnated
and coated fabrics, except rubberized
Glue and gelatin
Printing ink
Carbon black
Carbon and graphite products
Semiconductor (solid state) and related
devices
Storage batteries
Primary batteries (dry and wet)
Part of photographic equipment and
supplies: sensitized film, paper, cloth and
plates, and prepared photographic
chemicals for use therewith
Lead pencils, crayons, and artists' materials
Carbon paper and inked ribbons
Hard surface floor coverings
Other manufacturers

VI. Meat and Poultry Packing Plants
   A. Brine lines
   B. Caustic cleaning lines
   C. Auxiliary water lines
   D. Blood lines
   E. Viscera disposal
   F. Hide rendering and tanning lines
VII. Beverage Plants
   A. Distilled and deionized water lines
   B. Auxiliary water lines

Fig. 7-1. PVC ball valves used to unload chemicals at a railroad facility. (Courtesy of R&G Sloane.)

C. Caustic cleaning lines for vats and tanks
D. Carbon dioxide injection lines
E. Whiskey, liquors, beer, and soda lines

PLATING

The automotive, aircraft, electrotyping, and canning industries use thermoplastic piping wherever possible in their plating processes. Plastics are a natural in this market, since almost every metal-salt plating solution can be handled easily, including, but not limited to, brass, cad-

Fig. 7-2. PVC, CPVC, and Penton piping used in a catalyst room of a large chemical company. (Courtesy of Chemtrol Div. of Nibco.)

Fig. 7-3. Natural PVDF sanitary piping system used to handle high-purity water at a major pharmaceutical company. (Courtesy of Sani-Tech, Inc.)

mium, chrome, copper, gold, lead, nickel, rhodium, silver, tin, and zinc. Other areas in plating in which plastics are being used are in transfer feed and rinse lines and for most sulfuric and nitric acid lines. Secondary applications of plastic piping also are incorporated in the water-treatment and air-pollution-control systems required in most large plating facilities. (See Fig. 7-6–7-10.)

STEEL MILLS

Ironically, for over 25 years steel mills have been replacing steel piping with plastics. The mills re-

Fig. 7-4. PVDF piping system handling 30% nitric acid. (Courtesy of +GF+ Plastic Systems, Inc.)

Fig. 7-5. Complex PVC piping system for cucumber pickle plant. (Courtesy of Chemtrol Div. of Nibco.)

alized that their manufacturing costs improved owing to reduced maintenance, lower material costs, and much longer life using plastic piping. Estimates range from 40,000 to 65,000 gal of water needed to provide 1 ton of finished steel. The large amount of water and acids used in steel produc-

Fig. 7-7. CPVC piping used in a plating application at 150° F (66° C). (Courtesy of B. F. Goodrich.)

tion allows plastic piping to be used in the following applications:

I. Coke Plants
   A. Distillation towers
   B. Chemical process tanks
   C. Stills
   D. Washers
   E. Scrubbers
   F. Condensers

Fig. 7-6. Pumping station for electrolyte in galvanizing line showing polypropylene and PVC plastic pipe and valves. (Courtesy of +GF+ Plastic Systems.)

Fig. 7-8. CPVC and PVC piping systems used in neutralization piping in a major steel plant. (Courtesy of PPS-Maryland.)

G. Electrical conduit
H. Cold water
I. Hot water
J. Wastewater

II. Galvanizing Lines
A. Chemical feed
B. Mill water
C. Washer water
D. Tank connection
E. Scrubbers
F. Wastewater troughs

III. Pickle Lines
A. Chemical feed
B. Mill water
C. Chilled water
D. Wastewater
E. Hood exhausts
F. Scrubbers
G. Tank piping

IV. Tin Lines
A. Hot and cold water
B. Chemical feed
C. Hoods and scrubbers
D. Tank cleaning
E. Spray rinsing
F. Troughs

V. Waste Treatment Lines
A. Chemical feed
B. Acid waste
C. Acid storage tanks
D. Neutralization
E. Acid pumping pits
F. Battery storage

Fig. 7-9. Two inch CPVC piping being used in a pulp plant. (Courtesy of B. F. Goodrich.)

Fig. 7-10. Four inch CPVC expansion joint being used in a paper mill. (Courtesy of PPS-Maryland.)

PULP AND PAPER

Pulp and paper plants handle four types of media: liquids, steam, water, and stock. Except for steam, plastic piping handles most of the other fluids under 275° F (135° C) and 150 psi. Listed below are several applications in paper mills that plastic piping has successfully handled:

I. Rainwater Treatment
A. Sand filter and clarifier
B. Alum additive
C. Polymer addition

II. Power Plant
A. Demineralization and softening

III. Brown Stock Washing/Cleaning and Screening
A. Black liquor
B. Shower water
C. Stock

IV. Bleach Plant
A. Calcium hypochlorite
B. Sodium hypochlorite
C. Caustic soda
D. Chlorinated pulp
E. Hydrogen peroxide
F. Sodium chlorate
G. Dioxide plant spent acid
H. Sulfuric acid
I. White water

V. Stock Preparation
A. Alum
B. Clay slurry
C. Dyes

D. Calcium carbonate slurry
E. Pulp stock
F. Rosin
G. Stock sampling
H. Showers
I. White water
VI. Papermaking
A. Alum
B. Clay slurry
C. Dyes
D. Starch
E. White water
F. Size
VII. Recovery
A. Black liquor
B. Green liquor
C. Lime slurry
D. Soap
E. Tall oil
F. White liquor

ELECTRONICS

The manufacturers of solid-state electronics products such as semiconductors, rectifiers, printed circuitry, and transistors demand as pure water as possible to clean their products and prevent contamination. Thermoplastics are the preferred materials for handling ultrapure water in which an ion exchange or demineralization system is employed. PVC, CPVC, polypropylene, PVDF, and other fluorocarbons are being used for ultrapure water distribution systems. Thermoplastics are also being incorporated for handling etching media such as sulfuric, nitric, hydrochloric, and hydrofluoric acids. The etchants are used in most semiconductor and printed circuitry plants. As in most chemical processes, the wastewater and air-handling systems required in the electronic industry use plastic piping throughout. (See Figs. 7-11 and 7-12.)

PHOTOGRAPHIC LABORATORIES

Photographic laboratories have been one of the greatest growth industries in the United States. All manufacturers of photographic process equipment and photographic chemicals specify and use thermoplastics. PVC, CPVC, and polypropylene piping are used for reservoirs, tanks, film rollers,

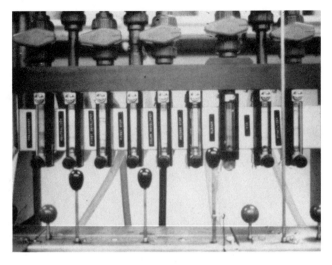

Fig. 7-11. Plastic-panel-mounted valving used in a printed circuit board manufacturing plant. (Courtesy of PPS-Maryland.)

"bottom rollers," and developer and stabilizer lines. There is also a particular use for PVC in bleaching tanks: In the bleaching operation there is a "Prussian Blue" buildup from oxidation that contaminates film; PVC minimizes this occurrence. Another specific use of PVC is as a dielectric shield near photocell tubes. (See Figs. 7-13 and 7-14.)

MINING/OIL/GAS

Plastic pipe, itself a derivative of oil and natural gas, is used to produce oil and natural gas

Fig. 7-12. Printed circuit board production using PVC manual and electrically activated valving. (Courtesy of +GF+ Plastic Systems.)

Fig. 7-13. Extensive PVC piping systems in a photographic operation. (Courtesy of R&G Sloane.)

more economically. Plastics have successfully been applied in handling all crudes, salt water and natural gases. In many oil and gas fields, paraffin buildup in metal lines carrying crude oil and gases has caused clogging of the lines. The petroleum industry now prefers to use plastic piping. Most commercial gas companies today use millions of feet of plastics in natural gas distribution. Polyethylene piping, colored beige or orange, is the preferred material for this application.

Mining has some of the "dirtiest" processes of any major industry. The most popular mining use of thermoplastics is in ore leaching, where the ore is treated with dilute sulfuric acid or sulfides and then with ferric sulfate solutions. The leaching is done in place, in tanks, or on pads where crushed ore is treated. RTRP piping and tanks, PVC, CPVC, ABS, and polyethylene piping are used in many of the leaching process stages. Another leaching-type application is when sulfuric acid percolates through to subterranean water. The resultant mixture of the water and acid cannot be handled successfully by metals. The uranium mining industry also uses miles of PVC and CPVC piping in their well casings and process lines. The movement of slurries is another excellent use of plastics. The abrasion resistance of PVC, ABS, polyethylene and RTRP piping offers much longer life than metal piping. Still another use of plastics in the mining industry is in the handling of acids and other sulfuric impurities in the air. Many smelters (especially copper) are being forced, and rightfully so, to emit minimum contaminators in the air from their stack gases. The by-product of this sulfur removal in many cases is sulfuric acid, which as stated before is handled easily by plastics. (See Figs. 7-15, 7-16, and 7-17.)

Fig. 7-14. Photographic film manufacturing facility showing polypropylene pipe fittings and valves. (Courtesy of +GF+ Plastic Systems.)

Fig. 7-15. Oil field piping using RTRP with fast mechanical joining system. (Courtesy of Smith Fiberglass Systems.)

Fig. 7-16. Oil field waste facility using 2–12-in. RTRP as disposal lines. (Courtesy of Smith Fiberglass Systems.)

Fig. 7-17. UL listed RTRP underground systems transferring petroleum products. (Courtesy of Smith Fiberglass Systems.)

MARINE APPLICATIONS

Shipbuilding, marinas, fish hatcheries, marine research, and amusement/theme parks are using significant amounts of plastic piping. From advanced atomic submarines to shrimp boats, thermoplastic piping is used on board ships for water and brine applications. The advantages of plastics in shipbuilding are:

1. No corrosion by salt water.

2. Light weight.

3. No corrosion with CO_2 impregnated brine solution.

4. Easily prefabricated in modules.

Many of today's modern tuna boats are equipped with PVC and CPVC in the hold cooling system as well as portable water washdown lines. Also, since space is at a premium, plastic piping can be used in cramped manways. Traditionally, crab and shrimp boats used freshwater ice to cool their catches. Owing to limited cooling capability and bacteria growth affecting the catch, the maximum time away from port was 4 days. However, by introducing carbon dioxide into brine, holds can be refrigerated so that crab and shrimp boats can remain at sea longer.

Dozens of applications use plastics in marinas, marine biology research, and fish hatcheries, such as in fish feed lines, waste disposal lines, and conduit and recreation lines in holding tanks. San Francisco, Boston, and Chicago aquariums all share something in common: the preponderance of thermoplastic and thermoset piping systems for freshwater and salt water lines. Plastics have proven to be the premier piping material for aquariums for the last 9 years. With amusement and theme parks such as DisneyLand, Disney-World, and Six Flags growing in numbers, PVC and RTRP piping are being used extensively in air conditioning systems, water treatment lines, waste lines as well as water supply lines for the many water attractions. (See Figs. 7-18 and 7-19.)

POWER PLANTS

The massive water requirements of power plants and, consequently, their enormous use of piping have naturally made an impact on the plastic pip-

Fig. 7-18. Aquarium RTR piping systems using 12-in. and smaller sizes. (Courtesy of Smith Fiberglass Systems.)

Fig. 7-20. Bottom and fly ash slurries from a power plant being carried by 10-in. ceramic-impregnated RTRP. (Courtesy of Smith Fiberglass Systems.)

ing industry. RTRP and polyethylene piping, in the largest diameters ever made in this country, are being used for intake and condenser water lines for the electric generating industry. Secondary water lines as well as scrubber piping and other water treatment lines are all using plastic piping in large quantities. Perhaps one of the most severe piping applications is the handling of fly and bottom ash slurries in coal-burning power plants. When transporting these ash slurries, ceramic-lined RTRP, polyethylene, and other plastics have outlived metal piping by thousands of hours. (See Figs. 7-20 and 7-21.)

SEWAGE TREATMENT

Sewage treatment is and will continue to be a growth industry throughout the world. Whether in primary or secondary treatment phases, plastics are used throughout a sewage treatment plant. (See Figs. 7-22–7-25.) Listed below are some of the areas of sewage-treatment applications plastic piping is successfully handling:

I. Settling Tanks
   A. Influent channel lines
   B. Effluent channel lines
   C. Scum collector lines
   D. Sludge withdrawal lines

Fig. 7-19. Dredging line using 10-in. polypropylene pipe. (Courtesy of Phillips Driscopipe.)

Fig. 7-21. Bottom ash slurry piping for a power plant utilizing 10-in. ceramic-impregnated RTRP. (Courtesy of Smith Fiberglass Systems.)

Fig. 7-22. Wastewater treatment system handling 30% HCl and 20% NaOH; the system uses RTRP and PVDF diaphragm valves. (Courtesy of +GF+ Plastic Systems.)

II. Trickling Filters
    A. Feed pipe
    B. Effluent channel lines
III. Sludge Treatment
    A. Sludge feed lines
    B. Water lines
    C. Chlorine feed lines

Fig. 7-23. Gas-flow-measurement systems in a sewage-treatment plant using PVC flanged ball valves. (Courtesy of +GF+ Plastic Systems.)

Fig. 7-24. PVC ball valves used with metal piping at a sewage-treatment plant. (Courtesy of R&G Sloane Manufacturing.)

    D. Sodium hypochlorite lines
IV. Sludge Treatment Disposal

WATER TREATMENT

The need for clean water is not geographically limited. No matter where an industry is located, its need for clean water is a constant problem. (See Figs. 7-26–7-29.) Plastics, which are inert to most chemical action, are particularly useful in water-treatment plants and may be successfully used in the following applications:

Chlorination

Sodium hydroxide

Fig. 7-25. One hundred and twenty inch polyethylene spirally wound pipe used in a large wastewater treatment plant. (Courtesy of Spirolite Corp.)

Fig. 7-26. Chlorine dioxide generation for sterilization of drinking water using PVC piping systems. (Courtesy of +GF+ Plastic Systems.)

Caustic soda

Sodium hypochlorite lines

Drain-off lines

Aeration equipment

Filtration lines

Ventilation ducting

Wash lines

Miscellaneous piping

Digestors

Plain water lines

Calgon chemical equipment for wells

Diatomaceous earth slurry lines

Calcium oxide lines

Potassium permanganate system

Powdered-activated carbon slurries

Fluoride lines

Alum lines

Chlorine feed lines

PLUMBING APPLICATIONS

By far the largest usage of plastics piping in the United States is in the plumbing markets. (See Fig. 7-30.) The above- and below ground usage of

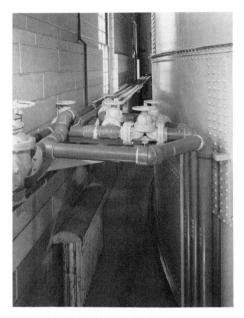

Fig. 7-27. PVC piping and metal valves used in a water-treatment facility. (Courtesy of Chemtrol Div. of Nibco.)

mostly thermoplastics has replaced many metal piping systems and has been approved by such code bodies as:

Building Officials and Code Administrators International

Fig. 7-28. PVC and polypropylene piping in a concrete walkway used in a water-filtration process. (Courtesy of Chemtrol Div. of Nibco.)

Fig. 7-29. CPVC and PVC piping in a wastewater-treatment plant. (Courtesy of Chemtrol Div. of Nibco.)

International Association of Plumbing and Mechanical Officials

International Conference of Building Officials

National Association of Plumbing, Heating, Cooling Contractors/American Society of Plumbing Engineers

Southern Building Code Congress International

Fig. 7-30. Sanitary sewer slip lining using 36-in. polyethylene piping. (Courtesy of Phillips Driscopipe.)

Plastic piping is also an approved material in U.S. government building projects as directed by the U.S. Department of Housing and Urban Development. Listed below are the applications and plastic material used for each plumbing use:

Plumbing Applications	Plastic Material						
	PVC	CPVC	SR	ABS	PE	PB	PP
Drain, water, and vent (DWV)	X			X			X
Hot/cold water distribution		X			X		
Outside sewers and drains	X		X	X	X	X	
Septic disposal/drainage	X		X	X			
Tubular waste	X			X			X
Water piping	X	X		X	X	X	

HEATING/AIR CONDITIONING/ REFRIGERATION

The market for plastic piping in commercial air conditioning systems is extremely large. Thousands of feet of RTRP, PVC, and CPVC piping are very commonplace in central air conditioning systems of hospitals, universities, apartment complexes, commercial buildings, and office areas of industrial manufacturers. (See Figs. 7-31, 7-32, and 7-33.)

The wide acceptance of plastics in the heating and air conditioning industry was gained initially in condensate return lines. Plastics are inert to the halogenated hydrocarbons that may exist in condensate return lines, whereas metal piping can be

Fig. 7-31. RTRP header system for cooling hot forging equipment. (Courtesy of Smith Fiberglass Systems.)

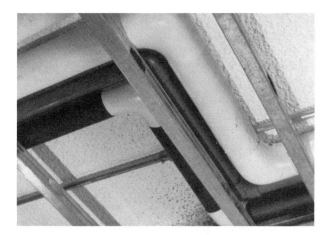

Fig. 7-32. CPVC pipe used in a hotel's fire sprinkler system. (Courtesy of R&G Sloane Manufacturing.)

corroded by them. Chilled water and condenser water lines were the next air conditioning services using plastic piping. PVC piping is increasingly being used in chilled water lines owing to its economical installed cost over copper (at least a 50–60% cost savings). Plastics are also used widely in refrigeration lines, because of their lower coefficient of thermal conductivity compared to metal piping. For example, PVC has a coefficient of thermal conductivity of 0.7 versus 2352 (Btu/hr/ft^2/°F/in.) of copper. This property of plastic piping results in less or even no piping insulation required compared to metal piping. Another excellent use of plastics is for the refrigeration process in ice skating rinks. Ultra-high-density high-molecular-weight polyethylene piping is being successfully used in dozens of ice rinks in the United States.

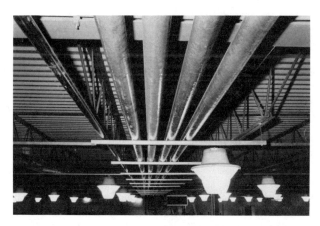

Fig. 7-33. RTRP hot and chilled water lines for plant heating and cooling. (Courtesy of Smith Fiberglass Systems.)

INSTITUTIONAL FACILITIES

Hospitals and school complexes are large users of plastic piping. These institutional facilities use PVC, CPVC, polyethylene, and RTRP in many chemical and corrosive services. Acid-waste drainage lines for chemistry, physics, and hospital laboratories use polypropylene. It is the most specified material for these applications. Other applications are PVC vacuum lines; CPVC for handling hot demineralized or deionized water; PVC for plumbing, air conditioning vent lines, irrigation systems; and RTRP in low-temperature-steam condensation lines. (See Figs. 7-34 and 7-35.)

Fig. 7-34. Polypropylene acid-waste drainage pipe in a college laboratory. (Courtesy of R&G Sloane Manufacturing.)

Fig. 7-35. Twelve-inch PVC gasketed pipe to be used for drainage at a large building site. (Courtesy of Uni-bell PVC Pipe Association.)

HEAVY CONSTRUCTION

The diversity and ingenuity of plastic piping for construction applications are both interesting and unusual. Plastic piping has been used as concrete forms, storage containers for other corrodable piping and electronic equipment, surge arrestors, drainage troughs, fence barriers, stands for holding warning signals, and dozens of other nonpipe applications. In construction of large dams, bridges, and tunnels, plastic piping is used to supply and or drain an area or site of water. Also water required for semihydraulic or hydraulic fill dams and for the preparation of concrete is easily handled by plastic piping. Transmitting of waterway dredging material is also a natural with thermoplastic and thermoset piping material. An important application of plastic piping is the pumping of water from the depths of the ocean to be used as a source of energy. Owing to the low temperature of the water, it is possible to condense moisture from the air in large volumes to help produce generation of electricity. After the cold water is warmed, it can be drained into collecting ponds in which various forms of marine life may be grown and harvested. Another very large usage of plastic piping (especially large diameters) is for drainage lines of roadways, bridges, and other overhead construction projects. (See Figs. 7-36 and 7-37.)

Fig. 7-36. Six-inch PVC drainage piping for a large commercial construction site. (Courtesy of R&G Sloane Manufacturing.)

Fig. 7-37. Storm drainage system using 24-in. polyethylene pipe. (Courtesy of Phillips Driscopipe.)

Appendix

GLOSSARY OF PLASTIC AND PIPING TECHNOLOGY

Acceptance test: An investigation performed on an individual lot of a previously qualified product, by, or under the observation of, the purchaser to establish conformity with a purchase agreement.

Acid vent: A pipe venting an acid waste system.

Acid waste: A pipe that conveys liquid waste matter having a pH of 6.9 or less.

Acme thread: A screw thread, the profile of which is between the square and V threads, used extensively for feed screws. The included angle between the flanks of the thread is 29° as compared to 60° for the Unified thread.

Acrylonitrile–butadiene–styrene (ABS) pipe and fitting plastics: Plastics containing polymers and/or blends of polymers, in which the minimum butadiene content is 6%, the minimum acrylonitrile content is 15%, the minimum styrene and/or substituted styrene content is 15%, and the maximum content of all other monomers is not more than 5%; they also contain lubricants, stabilizers, and colorants.

Adapter fitting: (1) Any of various fittings designed to mate or to fit to each other two pipes or fittings that are different in design, when connecting the two together would otherwise not be possible. (2) A fitting that serves to connect two different tubes or pipes to each other, such as copper tube to iron pipe.

Adhesive: A substance capable of holding materials together by surface attachment.

Adhesive solvent: An adhesive having a volatile organic liquid as a vehicle; the volatile organic liquid evaporates leaving behind the adhesive. (*See also* Solvent cement.)

Aging: The effect of exposing plastics to the environment for a length of time. The specific effect and degree of aging depend on the moisture, temperature, and

composition of the environment in addition to the length of exposure.

Air vent: Small outlets for preventing gas entrapment.

Ambient temperature: The prevailing temperature in the immediate vicinity, or the temperature of the medium surrounding an object.

Anchor: A device used to fasten or secure pipes to the building or structure.

Angle of bend: In a pipe, the angle between radial lines from the beginning and end of the bend to the center.

Anhydrous: Free of water, especially water of crystallization.

Antioxidant: A compounding ingredient added to a plastic composition to retard the degradation of the plastics' properties caused by contact with oxygen (air), particularly at or exposure to high temperatures.

Artificial weathering: The exposure of plastics to cyclic laboratory conditions involving changes in temperature, relative humidity, and ultraviolet radiant energy, with or without direct water spray, in an attempt to produce changes in the material similar to those observed after long-term continuous outdoor exposure. (*Note:* The laboratory exposure conditions are usually intensified beyond those encountered in actual outdoor exposure in an attempt to achieve an accelerated effect. This definition does not involve exposure to special conditions such as ozone, salt spray, industrial gases, etc.)

A-stage: Initial or early stage in the reaction of some thermosetting resins, where the material is still soluble in certain liquids and fusible; referred to as *resol*.

Autoclave: A closed vessel or reactor where chemical reactions take place under pressure.

Backfill: That portion of the trench excavation which is replaced after the buried pipe line has been laid with the material above the pipe (up to the original earth line).

Ball check valve: A device used to stop the flow of media in one direction while allowing flow in the opposite direction. The closure member used is spherical or ball shaped.

Ball valve: A valve with a ball-shaped disk with a hole through the center of the ball, providing straight-through flow. A quarter-turn of the handle fully opens or closes the valve for quick shut off.

Baumé gravity: Arbitrary scale for measuring the density of liquids; the unit used is the "Baumé" (Be) degree. The scale uses an inverse ratio of the specific gravity (sp. gr.) scale:

$$\text{sp. gr.} = \frac{140}{130 + \text{Be degree}}$$

(for liquids *lighter* than water)

$$\text{sp. gr.} = \frac{145}{145 - \text{Be degree}}$$

(for liquids *heavier* than water)

This permits the translation of Baumé gravity to specific gravity. For instance, when floated in pure water, the Baumé hydrometer indicates 10° Be, while the specific gravity scale reads 1.00. The Baumé scale is employed by the U.S. National Bureau of Standards for all liquids except oils.

Beam loading: The application of a load to a pipe between two points of support; it is usually expressed in pounds and the distance between the centers of supports.

Bell and spigot joints: One side of the fitting or pipe is belled or socket; the other end is plain-ended pipe.

Bell end: The enlarged portion of a pipe that resembles the socket portion of a fitting and that is intended to make a joint by inserting a piece of pipe into it. Joining may be accomplished by solvent cements, adhesives, or mechanical techniques.

Binder: The part of adhesive composition responsible for adhesive forces.

Blister: The elevated part of the surface of a plastic caused by trapped air, moisture, solvent; it can be caused by insufficient adhesive, inadequate curing time, or excessive temperature or pressure.

Block valve: Valve used for isolating equipment.

Bond: The attachment at the interface or exposed surfaces between an adhesive and an adherent; to attach materials together with adhesives.

Branch: Any part of a piping system other than a main, riser, or stack.

Branch interval: A length of soil or waste stack corresponding in general to a story height—but in no case less than 8 ft (2.4 m)—within which the horizontal branches from one floor or story of a building are connected to the stack.

Branch tee: A tee having one side branch.

Branch vent: A vent connecting one or more individual vents with a vent stack or stack vent.

British thermal unit (Btu): The quantity of energy needed to heat one pound of water from 59° F to 60° F at a standard barometric pressure; 1 Btu = 0.252 kcal = 0.000293 kWh.

B-stage: Intermediate-stage reaction step for various thermosetting resins. During this stage, the material swells when in contact with certain liquids and becomes very soft when heat is applied. The material may not dissolve or fuse entirely. Resin in this stage is referred to as *resitol.*

Bubble tight: The condition of a valve seat that, when closed, prohibits the leakage of visible bubbles.

Bulk density: Density of a molding material in loose form, such as granular or nodular, in units of g/cm^3 or lb/ft^3.

Bulkhead fitting: A fitting fixed to a vessel wall that allows fluids to flow through the wall and adapts pipe or tubing to the vessel. One end is usually smooth (inside vessel), and the other end is usually threaded.

Bull head tee: A tee, the branch of which is larger than the run.

Burst pressure: The pressure that can be applied slowly to a valve, fitting, or pipe at room temperature for 30 sec without causing rupture.

Burst strength: The internal pressure required to break a pipe or fitting. This pressure will vary with the rate of build up of the pressure and the time during which the pressure is maintained.

Bushing: A fitting used to connect a pipe to a female fitting of a larger size.

Butt fusion: A method of joining thermoplastic pipes wherein the ends of the two pieces to be joined are heated to the molten state and then quickly pressed together.

Butt weld joint: A welded pipe joint made with the ends of the two pipes butting each other.

Butterfly valve: A device deriving its name from the winglike action of the disk, which operates at right angles to the flow. The disk impinges on the resilient seal with low operating torque.

By-pass: An auxiliary loop in a pipeline that diverts flow around a valve or other piece of equipment.

By-pass valve: Valve by which the flow or liquid or gas in a system may be directed past some part of the system through which it normally flows (e.g., an oil filter in a lubrication system).

Capacity: The maximum or minimum flow obtainable under given conditions of media, temperature, pressure, velocity, etc. Also, the volume of media that may be stored in a container or receptacle.

Capillary: The action by which the surface of a liquid, where it is in contact with a solid, is elevated or depressed depending on the relative attraction of the molecules of the liquid for each other and for those of the solid.

Case harden: Process of hardening the surface of a piece of steel to a relatively shallow depth.

Catalyst: Material used to activate resins to promote hardening. For polyesters, organic peroxides are primarily used. For expoxies, amines and anhydrides are used.

Cement: A dispersion or "solution" of unvulcanized rubber or a plastic in a volatile solvent. This meaning is peculiar to the plastics and rubber industries and may or may not be an adhesive composition. (*See also* Adhesive and Solvent cement.)

Centipoise: Unit of absolute viscosity; it equals one hundredth of a poise.

Centistoke: Unit of kinematic viscosity; it equals one hundredth of a stoke. Kinematic viscosity in centistokes multiplied by the specific gravity equals absolute viscosity in centipoises.

Centrifugal casting: Process in which tubular products are fabricated by applying resin and glass-strand reinforcement to the inside of a mold that is rotated and heated. The process polymerizes the resin system.

Chase: A recess in a wall in which pipes can be run.

Check valve: Device that permits flow in only one direction in a pipeline.

Chemical waste system: Piping that conveys corrosive or harmful industrial, chemical, or processed wastes to the drainage system.

Chemical resistance: (1) The effect of specific chemicals on the properties of plastic piping with respect to concentration, temperature, and time of exposure. (2) The ability of a specific plastic pipe to render service for a useful period in the transport of a specific chemical at a specified concentration and temperature.

Circuit vent: A branch vent that serves two or more traps and extends from in front of the last fixture connection of a horizontal branch to the vent stack.

Cleaner: Medium strength organic solvent such as methylethyl ketone used to remove foreign matter from plastic pipe and fitting joint surfaces.

Cleanout: A plug or cover joined to an opening in a pipe that can be removed to clean or examine the interior of the pipe.

Close nipple: A nipple twice as long as a standard pipe thread.

Coefficient of expansion: The increase in unit length, area, or volume for a unit rise in temperature.

Cold flow (*see* Creep).

Cold molding: The fashioning of an unheated mixture in a mold under pressure. The article is then heated to effect curing.

Cold pressing: Bonding process in which an assembly is subjected to pressure without applying heat.

Combined waste and vent system: A specially designed system of waste piping, embodying the horizontal wet venting of one or more floor sinks or floor drains by means of a common waste and vent pipe, adequately sized to provide free movement of air above the flow line of the drain.

Combustion: Burning or rapid oxidation.

Common vent: A vent that connects at the junction of two fixture drains and serves as a vent for both fixtures. Also, known as a dual vent.

Companion flange: A pipe flange used to connect with another pipe flange or with a flanged valve or fitting. It is attached to the pipe by threads, welding, or other methods and differs from a flange that is an integral part of a pipe or fitting.

Compound: The admixture of a polymer or polymers with other ingredients such as fillers, softeners, plasticizers, catalysts, pigments, dyes, curing agents, stabilizers, or antioxidants.

Compression fitting: A fitting designed to join a pipe or tube by means of pressure or friction.

Compression joint: A multipiece joint with cup-shaped threaded nuts, which, when tightened, compress tapered sleeves so that they form a tight joint on the periphery of the tubing they connect.

Compression strength: The crushing load at failure of an item divided by the number of square inches of resisting surface. (Figures are given in thousands of pounds.)

Condensate: Water that has liquefied from steam.

Contact molding: Process in which layers of resin-impregnated fabrics are built up one layer at a time onto the mold surface forming the product. Little or no pressure is required for laminate curing.

Continuous vent: A vent that is a continuation of the drain to which it connects.

Continuous waste: A drain from two or three fixtures connected to a single trap.

Control valve: Variable opening valve used with a control instrument to maintain a predetermined flow rate, pressure, temperature, or level. The valve can be electric, electrohydraulic, or air operated.

Copolymer: Formed from two or more monomers. (*See also* Polymer.)

Crazing: Tiny cracks that develop on a laminate's surface; these cracks are caused by mechanical or thermal stress.

Creep: The time-dependent part of strain resulting from stress; that is, the dimensional change with respect to time caused by the application of load over and above the elastic deformation.

Cross: A pipe fitting with four branches in pairs, each pair on one axis, and the axes at right angles.

Cross-linking: Generation of chemical linkages between long-chain molecules; cross-linking can be compared to two straight chains joined together by links. The rigidity of the material increases with the number of links. The function of a monomer is to provide these links.

Cross-over: A fitting with a double offset—shaped like the letter "U" with the ends turned out—used to pass the flow of one pipe past another when the pipes are in the same plane.

Crown: That part of the trap in which the direction of flow is changed from upward to downward.

Crown vent: A vent pipe connected at the topmost point in the crown of a trap.

C-stage: Final reaction stage of various thermosetting resins. In this stage material is insoluble and infusible. Resin in fully cured thermosetting molding is in this stage and is referred to as *resite.*

Cure: Process in which the addition of heat or catalyst or both, with or without pressure, causes the physical properties of the plastic to change through a chemical reaction, which may be condensation, polymerization, or addition reactions.

C-veil: Thin, nonwoven fabric composed of randomly oriented and adhered glass fibers of a chemically resistant glass mixture.

Dead end: A branch leading from a soil, waste, or vent pipe; building (house) drain; or building (house) sewer that is terminated at a developed distance of 2 ft (0.6 m) or more by means of a plug or other closed fitting.

Deflection temperature: The temperature at which a specimen will deflect a given distance at a given load under prescribed conditions of test (formerly called heat distortion).

Degradation: Deleterious change in a plastic's chemical structure.

Delamination: Separation of a laminate's layers.

Density: Mass of a fluid per unit volume.

Deterioration: Permanent adverse change in the physical properties of a plastic.

Developed length: The length along the center line of the pipe and fittings.

Dew point: The temperature at which liquid first condenses when a vapor is cooled.

Diameter: Unless specifically stated, the nominal diameter as designated commercially.

Diaphragm: A flexible disk used to separate the control medium from the controlled medium; it actuates the valve stem.

Diaphragm valve: A valve used for controlling flow by a flexible elastometric disk.

Diffusion: The movement of a material, such as a gas or liquid, in the body of a plastic. If the gas or liquid is absorbed on one side of a piece of plastic and given off on the other side, the phenomenon is called permeability. Diffusion and permeability are not due to holes or pores in the plastic, but are caused and controlled by chemical mechanisms.

Dimension ratio: The diameter of a pipe divided by the wall thickness. Each pipe can have two dimension ratios depending on whether the pipe's outside or inside diameter is used. In practice, the outside diameter is used if the standards requirement and manufacturing control are based on this diameter. The inside diameter is used when this measurement is the controlling one.

Disk: The part of a valve that actually closes off the flow.

Dispersion: Heterogeneous mixture in which finely divided material is suspended in the matrix of another material; as in the distribution of finely divided solids in a liquid or a solid (e.g., pigments or fillers).

Displacement: The volume or weight of a fluid, such as water, displaced by a floating body.

Doping: Coating a mandrel or mold with a material that prevents the finished product from sticking to it.

Double offset: Two changes of direction installed in succession or series in continuous pipe.

Double ported valve: A valve having two ports to overcome line pressure imbalance.

Double sweep tee: A tee made with "easy" (long-radius) curves between body and branch.

Drain: Any pipe that carries wastewater or water-borne wastes in a building drainage system.

Drainage fitting: A type of fitting used for draining fluid from pipes. The fitting makes a smooth and continuous interior surface for the piping system.

Dry-blend: A free-flowing dry compound prepared without fluxing or addition of solvent.

Dry bulb temperature: The temperature of air as measured by an ordinary thermometer.

Dry spot: Incompleted area on laminated plastics; the region in which the interlayer and glass are not bonded.

Dual vent (*see* Common vent).

Durometer hardness: A material's hardness as measured by the Shore Durometer.

Elasticity: Property of materials that tends to retain or recover their original shape and size after undergoing deformation.

Elastomer: A material under ambient conditions that can be stretched and, upon release of the applied stress, returns with force to approximately its original size and shape.

Elevated temperature testing: Tests on plastic pipe above 23° C (73° F).

Elbow (Ell): A fitting that makes an angle between adjacent pipes. The angle is 90°, unless another angle is specified.

End connection: A reference to the method of connecting the parts of a piping system (i.e., threaded, flanged, butt-weld, socket weld).

Environmental stress cracking: Cracks that develop when the material is subjected to stress in the presence of specific chemicals.

Epoxy plastics: Group of plastics composed of resins produced by reactions of epoxides or oxiranes with compounds such as amines, phenols, alcohols, carboxylic acids, acid anhydrides, and unsaturated compounds.

Erosion: The gradual destruction of a material by the abrasive action of liquids, gases, solids, or mixtures of these materials.

Expansion joint: Joint used in the connection of long lines of pipe; the expansion joint contains a bellows or telescopelike section to absorb the thrust or stress resulting from linear expansion or contraction of the line owing to changes in temperature or to accidental forces.

Expansion loop: A large radius bend in a pipe line to absorb longitudinal thermal expansion in the line.

Extender: A material that, when added to an adhesive, reduces the amount of primary binder necessary.

Extrusion: Method of processing plastic in a continuous or extended form by forcing the heat-softened plastic through an opening shaped like the cross section of the finished product.

Fabricate: Method of forming a plastic into a finished article by machining, drawing, and similar operations.

Face-to-face dimensions: The dimensions from the face of the inlet port to the face of the outlet port of a valve or fitting.

Female thread: Internal thread in pipe fittings, valves, etc., for making screwed connections.

Fiber stress: The unit stress, usually in pounds per square inch (psi), in a piece of material that is subjected to an external load.

Filament winding: Process in which continuous strands of roving or roving tape are wound, at a specified pitch and tension, onto the outside surface of a mandrel. Roving is saturated with liquid resin or is preimpregnated with partially cured resin. Application of heat may be required to promote polymerization.

Filler: A relatively inert material added to a plastic to modify its strength, permanence, working properties, or other qualities, or to lower costs.

Filter: A device through which fluid is passed to separate contaminates.

Filter element: A porous device that performs filtration.

Finishing: Removal of any defects from the surfaces of plastic products.

Flammability: The time a specimen will support a flame after having been exposed to it for a given period of time.

Flange: In pipe work, a ring plate on the end of a pipe at right angles to the end of the pipe and provided with holes for bolts to allow fastening the pipe to a similarly equipped adjoining pipe. The resulting joint is called a flanged joint.

Flanged ends: A valve or fitting having flanges for joining to other piping elements. Flanged ends can be plain faced, raised face, large male and female, large tongue and groove, small tongue and groove, or ring joint.

Flange faces: Pipe flanges that have the entire surface of the flange faced straight across and use either a full face or ring gasket.

Flash point: Temperature at which enough of a material is vaporized to produce a flash of burning vapor.

Flexural strength: The pressure in pounds necessary to break a given sample when the pressure is applied to the center of the sample that has been supported at its ends.

Flow coefficient or C: Valve coefficient of flow representing the flow rate of water in gallons per minute that will produce a 1 psi pressure drop through the valve.

Fluorocarbon resins: Material made by the polymerization of monomers composed only of carbon and fluorine.

Foot valve: Check valve located at the inlet end of the suction line at a pump that allows the pump to remain full of liquid when not in service.

Forming: A process in which the shape of plastic pieces such as sheets, rods, or tubes is changed to a desired configuration. (*Note:* The use of the term "forming" in plastics technology does not include such operations as molding, casting, or extrusion, in which shapes or pieces are made from molding materials or liquids.)

Full port valve: A valve in which the resistance to flow, in the open position, is equal to an equivalent length of pipe.

Fungi resistance: The ability of plastic pipe to withstand fungi growth or their metabolic products or both under normal conditions of service or laboratory tests simulating such conditions.

Furane plastics: Group of plastics composed of resins in which the furane ring is an integral portion of the polymer chain; made from polymerization or polycondensation of furfural, furfural alcohol, and other compounds containing furane rings; also formed by reaction of furane compounds with an equal or less weight of other compounds.

Fusion point: Temperature at which solid and liquid states of a substance can exist together in equilibrium (also called melting or freezing points).

Gate valve: Valve with a sliding blank that opens to the complete cross section of the line; used for complete opening or complete shutoff of the flow in pipes. It is not used for throttling or control.

Gel: State at which resin exists before becoming a hard solid. Resin material has the consistency of a gelatin in this state; initial jellylike solid phase that develops during the formation of a resin from a liquid.

Gel coat: Specially formulated polyester resin that is pigmented and contains fillers. It provides a smooth, pore-free surface for the plastic article.

Gel point: Stage at which liquid begins to show pseudoelastic properties.

Gelation: Formation of a gel.

Glass: Inorganic product of a fusion reaction. Material forms upon cooling to a rigid state without undergoing crystallization. Glass is typically hard and brittle and will fracture conchoidally (radially).

Glass transition temperature: The range of temperatures in which a plastic changes from a rigid to a soft state. (*Note:* values will depend on the method of test. It is sometimes referred to as softening point.)

Globe valve: Valve used for throttling that does not have a straight-through opening.

Grade: The slope or fall of a line of pipe in reference to a horizontal plane. In drainage, it is expressed as the fall in a fraction of an inch or percentage slope per foot (mm/m) length of pipe.

Hand lay up molding (*see* Contact molding).

Hardener: Compound or mixture that, when added to an adhesive, promotes curing.

Haunching: Area from the bedding to the spring line of pipe. Provides the majority of load carrying of un-

derground pipe and also provides side support for flexible and rigid pipe.

Heat capacity: The quantity of heat required to raise the temperature of a given mass by 1°. This quantity is based on either 1 mole or a unit mass of material.

Heat of fusion: Heat needed to change a quantity of solid to a liquid, without a change in temperature.

Heat joining: Making a pipe joint by heating the edges of the parts to be joined so that they fuse and become essentially one piece with or without additional material.

Heat treat: Refers to annealing, hardening, and tempering of metals.

Hoop stress: The tensile stress, usually expressed in pounds per square inch (psi), in the circumferential orientation in the wall of the pipe when the pipe contains a gas or liquid under pressure.

Hot soils: Soils having a resistivity of less than 1000 ohm-cm; they are generally very corrosive to base steel.

Hydraulic gradient: The amount of inclination of a drainage line between the trap outlet and the vent connection, not exceeding one pipe diameter in this total length.

Hydrostatic design stress: The estimated maximum tensile stress in the wall of the pipe in the circumferential orientation owing to internal hydrostatic pressure that can be applied continuously with a high degree of certainty that failure of the pipe will not occur.

Hydrostatic strength (quick): The hoop stress calculated by means of the ISO equation at which the pipe breaks due to an internal pressure buildup, usually within 60–70 sec. (*See also* ISO equation.)

Impact, Izod: A specific type of impact test made with a pendulum-type machine. The specimens are molded or extruded, with a machined notch in the center. (*See also* Izod impact strength.)

Impact strength: Resistance, or mechanical energy absorbed by a plastic part, to shocks such as dropping and hard blows.

Impact, tup: A falling weight (tup) impact test developed specifically for pipe and fittings. There are several variables that can be selected.

Inhibitor: Material that retards chemical reaction or curing.

ISO equation: An equation showing the relations among stress, pressure, and dimensions in pipe, namely,

$$S = \frac{P(\mathrm{ID} + t)}{2t}$$

or

$$\frac{P(\mathrm{OD} - t)}{2t}$$

where

S = stress

P = pressure

ID = average inside diameter

OD = average outside diameter

t = minimum wall thickness

Isobaric process: A constant-pressure process.

Isometric process: A constant-volume process.

Isothermal process: A constant-temperature process.

Isotropic: Refers to materials whose properties are the same in all directions. Examples of isotropic materials are metals and glass mats.

Izod impact strength: The resistance a notched test specimen has to a sharp blow from a pendulum hammer. (*See also* Impact, Izod.)

Joint: The location at which two pieces of pipe or a pipe and a fitting are connected. The joint may be made by a mechanical device, such as threads or ring seals, or by heat fusion and cementing.

Laminate: Article fabricated by bonding together several layers of material or materials.

Laminated, cross: Laminate in which some of the layers of materials are oriented at right angles to the other layers. Orientation may be based on grain or strength direction considerations.

Laminated, parallel: Laminate in which all layers of materials are oriented parallel with respect to grain or strongest direction in tension.

Load factor: The percentage of the total connected fixture unit flow that is likely to occur at any point in the drainage system. Load factor represents the ratio of the probable load to the potential load and is determined by the average rates of flow of the various kinds of fixtures, the average frequency of use, the duration of flow during one use, and the number of fixtures installed.

Long-term burst: The internal pressure at which a pipe or fitting will break due to a constant internal pressure held for 100,000 hr (11.43 years).

Long-term hydrostatic strength: The estimated tensile stress in the wall of the pipe on the circumferential orientation (hoop stress) that when applied continuously will cause failure of the pipe at 100,000 hr (11.43 years). These strengths are usually obtained by extrapolation of log–log regression equations or plots.

Lubricant: A substance used to decrease the friction between solid faces, and sometimes used to improve processing characteristics of rubber or plastic compositions.

Modulus: The load in pounds per square inch, kilograms (force) per square centimeter, or, in SI, the mod-

ern metric system, megapascals (MPa) of initial cross-section area necessary to produce a stated percentage elongation; this value is used in the physical testing of rubber or plastics. It is a measure of stiffness, and is influenced by pigmentation, state of cure, quality of raw rubber, and other factors. The modulus at any given elongation is shown by stress–strain curve.

Molding, blow: Method of forming plastic articles by inflating masses of plastic material with compressed gas.

Molding, compression: Process of shaping plastic articles by placing material in a confining mold cavity and applying pressure and usually heat.

Molding, contact-pressure: Method of molding or laminating in which the pressure used is slightly greater than is necessary to bind materials together during the molding stage (the pressure generally less than 10 psi).

Molding, high-pressure: Molding or laminating with pressures in excess of 200 psi.

Molding, injection: Process of making plastic articles from powdered or granular plastics by fusing the material in a chamber under pressure with heat and forcing part of the mass into a cooler cavity where it solidifies; used primarily with thermoplastics.

Molding, low-pressure: Molding or laminating with pressures below 200 psi.

Molding, transfer: Process of molding plastic articles from powered, granular, or performed plastics by fusing the material in a chamber with heat and forcing the mass into a hot chamber for solidification. Used primarily with thermosetting plastics.

Monomer: Reactive material that is compatible with the basic resin. Tends to lower the viscosity of the resin. Simplest repeating structural unit of polymer.

Needle valve: Valve with a cone seat and needle-point plug to control small and accurate flows.

Nonrigid plastic: Plastic whose apparent modulus of elasticity is not greater than 10,000 psi at room temperature in accordance with the Standard Method of Test for Stiffness in Flexure of Plastics.

Nylon plastics: Group of plastics comprised of resins that are primarily long-chain synthetic polymeric amides. These resins have recurring amide groups as an integral part of the principal polymer chain.

Offset: A combination of pipe, pipes, and/or fittings that join two approximately parallel sections of the line of pipe.

Olefin plastics: Plastics based on resins made by the polymerization of olefins or copolymerization of olefins with other unsaturated compounds, the olefins being in greatest amount by weight. Polyethylene, polypropylene, and polybutylene are the most common olefin plastics encountered as pipe.

Outdoor exposure: Plastic pipe placed in service or stored so that it is not protected from the elements of normal weather conditions, i.e., the sun's rays, rain, air, and wind. Exposure to industrial and waste gases, chemicals, engine exhausts, etc., are not considered normal "outdoor exposure."

Permanence: The property of a plastic that describes its resistance to appreciable changes in characteristics with time and environment.

Permeability (*see* Diffusion).

Pitch: The amount of slope or grade given to horizontal piping and expressed in inches of vertically projected drop per foot (or mm/m) on a horizontally projected run of pipe.

Pipe stiffness: A measure of how a flexible conduit will behave under burial conditions.

Plastic (n): A material that contains as an essential ingredient an organic substance of large molecular weight, is solid in its finished state, and, at some state in its manufacture or in its processing into finished articles, can be shaped by flow.

Plastic (adj.): The adjective plastic indicates that the noun modified is made of, consists of, or pertains to plastic. (*Note 1:* This definition may be used as a separate meaning to the definitions contained in the dictionary for the adjective "plastic." *Note 2:* The plural form may be used to refer to two or more plastic materials, for example, plastics industry. However, when the intent is to distinguish "plastic products" from "wood products" or "glass products," the singular form should be used. As a general rule, if the adjective is to restrict the noun modified with respect to type of material, "plastic" should be used; if the adjective is to indicate that more than one type of plastic material is or may be involved, "plastics" is permissible.)

Plastic conduit: Plastic pipe or tubing used as an enclosure for electrical wiring.

Plastic pipe: A hollow cylinder of plastic material in which the wall thickness is usually small compared to the diameter and in which the inside and outside walls are essentially concentric. (*See also* Plastic tubing.)

Plastic, semirigid: Plastic having apparent modulus of elasticity in the range of 10,000–100,000 psi at 23° C, as determined by the Standard Method of Test for Stiffness in Flexure Plastics.

Plastic tubing: A particular size of plastic in which the outside diameter is essentially the same as that of copper tubing. (*See also* Plastic pipe.)

Plastic welding: Joining of finished plastic components by fusing materials either with or without the addition of plastic from another source.

Plasticate: Softening by heating or kneading.

Plasticity: Property of plastics that permits the material to undergo deformation permanently and continuously without rupture from a force that exceeds the yield value of the material.

Plasticize: Softening by adding a plasticizer.

Plasticizer: Material added to a plastic to increase its workability and flexibility. Plasticizers tend to lower the melt viscosity, the glass transition temperature, and/ or the elastic modulus.

Plug valve: Valve mainly used in gas service; consists of a rotating cylindrical plug in a cylindrical housing with an opening running through the plug.

Polybutylene: A polymer prepared by the polymerization of butene-1 as the sole monomer. (*See also* Polybutylene plastics.)

Polybutylene plastics: Plastics based on polymers made with butene-1 as essentially the sole monomer.

Polyethylene: A polymer prepared by the polymerization of ethylene as the sole monomer. (*See also* Polyethylene plastics.)

Polyethylene plastics: Plastics based on polymers made with ethylene as essentially the sole monomer. (*Note:* In common usage for this plastic, polyethylene plastic essentially means no less than 85% ethylene and no less than 95% total olefins.)

Polymer: A compound formed by the reaction of simple molecules having functional groups that permit their combination to proceed to high molecular weights under suitable conditions. Polymers may be formed by polymerization (addition polymer) or polycondensation (condensation polymer). When two or more monomers are involved, the product is called a copolymer.

Polymerization: A chemical reaction in which the molecules of a monomer are linked together to form large molecules whose molecular weight is a multiple of that of the original substance. When two or more monomers are involved, the process is called copolymerization or heteropolymerization.

Polyolefin: A polymer prepared by the polymerization of an olefin(s) as the sole monomer(s). (*See also* Polyolefin plastics and Olefin plastics.)

Polyolefin plastics: Plastics based on the polymers made with an olefin(s) as essentially the sole monomer(s).

Polypropylene: A polymer prepared by the polymerization of propylene as the sole monomer. (*See also* Polypropylene plastics and Propylene plastics.)

Polypropylene plastics: Plastics based on polymers made with propylene as essentially the sole monomer.

Polystyrene: A plastic based on a resin made by polymerization of styrene as the sole monomer. (*See also* Styrene plastics.) (*Note:* Polystyrene may contain minor proportions of lubricants, stabilizers, fillers, pigments, and dyes.)

Polyvinyl chloride: A resin prepared by the polymerization of vinyl chloride with or without the addition of small amounts of other monomers.

Polyvinyl chloride plastics: Plastics made by combining polyvinyl chloride with colorants, fillers, plasticizers, stabilizers, lubricants, other polymers, and other compounding ingredients. Not all of these modifiers are used in pipe compounds.

Pot life: Time period beginning once the resin is catalyzed and terminating when the material is no longer workable; working life.

Powder blend (*see* Dry-blend).

Preform: Coherent block of granular plastic molding compound or of fibrous mixture with or without resin. Prepared by sufficiently compressing material, forming a block that can be handled readily.

Prepolymer: An intermediate chemical structure between that of a monomer and the final resin.

Pressure: When expressed with reference to pipe, the force per unit area exerted by the medium in the pipe.

Pressure rating: The estimated maximum pressure that the medium in the pipe can exert continuously with a high degree of certainty that failure of the pipe will not occur.

Pressure tubing: Tubing used to convey fluids at elevated temperatures and/or pressures. Suitable for head applications, it is fabricated to exact outside diameter and decimal wall thickness in sizes ranging from $\frac{1}{2}$ to 6 in. outside diameter inclusive and to ASTM specifications.

Primer: Coating that is applied to a surface before application of an adhesive, enamel, etc.; its purpose is to improve bonding.

Promoted resin: Resin with an accelerator but no catalyst added.

Propylene plastics: Plastics based on resins made by the polymerization of propylene or copolymerization of propylene with one or more unsaturated compounds, the propylene being in greatest amount by weight.

Pump: Mechanical device for transporting liquids in pipelines; major types are centrifugal, reciprocating, turbine, rotary, and proportioning.

Qualification test: An investigation, independent of a procurement action, performed on a product to determine whether or not the product conforms to all requirements of the applicable specification. (*Note:* The examination is usually conducted by the agency responsible for the specification, the purchaser, or facility approved by the purchaser, at the request of the supplier seeking inclusion of his or her product on a qualified products list.)

Quick burst: The internal pressure required to burst a pipe or fitting due to an internal pressure buildup, usually within 60–90 sec.

Reinforced plastic: According to ASTM, those plastics having superior properties over those consisting of the base resin, owing to the presence of high-strength fillers embedded in the composition. Reinforced fillers are usually fibers, fabrics, beads, or mats made of fibers.

Relief valve: Safety device for automatic release of fluid at a predetermined pressure.

Resin: A solid, semisolid, or pseudosolid organic material that has an indefinite and often high molecular weight, exhibits a tendency to flow when subjected to stress, usually has a softening or melting range, and usually fractures conchoidally (radially).

Reworked material (thermoplastic): A plastic material that has been reprocessed, after having been previously processed by molding, extrusion, etc., in a fabricator's plant.

Riser: A water supply pipe that extends vertically one or more stories to transport water to fixtures or branches.

Roller: A serrated piece of aluminum used to work a plastic laminate. The purpose of the device is to compact a laminate and to break up larger air pockets to permit release of entrapped air.

Slide valve: Valve comprised of a body with a large sliding disk, usually in the horizontal plane, which is actuated by compressed air or a hydraulic cylinder.

Softening range: The range of temperature in which a plastic changes from a rigid to a soft state. (*Note:* Actual values will depend on the method of test. It is sometimes referred to as softening point.)

Solvent cement: In the plastic piping field, an adhesive that contains a solvent which dissolves or softens the surfaces being bonded so that the bonded assembly becomes essentially one piece of the same type of plastic.

Specific gravity: Weight of a unit of fluid volume.

Specimen: An individual piece or portion of a sample used to make a specific test. Specific tests usually require specimens of specific shape and dimensions.

Spool piece: A measured length of piping usually flanged on both ends.

Stabilizer: A compounding ingredient added to a plastic composition to retard possible degradation on exposure to high temperatures, particularly in processing. An antioxidant, for example, is a specific kind of stabilizer.

Stack: The vertical main of a system of soil, waste, or vent piping extending through one or more stories.

Standard dimension ratio (SDR): A selected series of numbers in which the dimension ratios are constants for all sizes of pipe.

Stiffness factor: A physical property of plastic pipe that indicates the degree of flexibility of the pipe when subjected to external loads.

Strain: The ratio of the amount of deformation to the length being deformed caused by the application of a load on a piece of material.

Roving: Bundle of continuous, untwisted glass fibers. Glass fibers are wound onto a roll called a "roving package."

Rubber: A material that is capable of recovering from large deformations quickly and forcibly. (*See also* Elastomer.)

Sample: A small part or portion of a plastic material or product intended to be representative of the whole.

Saran plastics: Plastics based on resins made by the polymerization of vinylidene chloride with other unsaturated compounds, the vinylidene chloride being in greatest amount by weight.

Schedule: A pipe size system (outside diameters and wall thickness) originated by the Iron Pipe Industry.

Self-extinguishing: The ability of a plastic to resist burning when the source of heat or flame that ignited it is removed.

Service factor: A factor that is used to reduce a strength value to obtain an engineering design stress. The factor may vary depending on the service desired and the properties of the pipe.

Set: To convert an adhesive into a fixed or hardened state by chemical or physical action, such as condensation, polymerization, oxidation, vulcanization, gelation, hydration, or evaporation of volatile constituents. (*See also* Cure.)

Shelf life: Period of time over which a material will remain usable during storage under specified conditions such as temperature and humidity.

Strength: The stress required to break, rupture, or cause a failure.

Stress crack: External or internal cracks in a plastic caused by tensile stresses less than that of the plastic's short-time mechanical strength. (*Note:* The development of such cracks is frequently accelerated by the environment to which the plastic is exposed. The stresses that cause cracking may be present internally or externally or may be combinations of these stresses.)

Street elbow: Pipe fitting with a male thread on one end and a female thread on the other end.

Styrene plastics: Plastics based on resins made by the polymerization of styrene or copolymerization of styrene with other unsaturated compounds, the styrene being in greatest amount by weight.

Styrene–rubber (SR) plastics: Compositions based on rubbers and styrene plastics, the styrene plastics being in greatest amount by weight.

Styrene–rubber (SR) pipe and fitting plastics: Plastics containing at least 50% styrene plastics combined with rubbers and other compounding materials, but not more than 15% acrylonitrile.

Sustained pressure test: A constant internal pressure test for 1000 hr.

Tank adapter (*see* Bulkhead fitting).

Temperature: A measure of the degree of hotness of a material detected most commonly with a liquid-in-glass thermometer.

Tensile strength: The capacity of a material to resist a force tending to stretch it. Ordinarily, the term is used to denote the force required to stretch a material to rupture, and is known variously as "breaking load," "breaking stress," "ultimate tensile strength," and sometimes erroneously as "breaking strain." In rubber and plastics testing, it is the load in pounds per square inch, kilograms per square centimeter, or newtons per square millimeter in modern SI metric, of original cross-sectional area, supported at the moment of rupture by a piece of test sample on being elongated.

Thermal conductance: Also called "conductance," it is the amount of heat transmitted by a material divided by the difference in temperature of the surfaces of the material. Where heat is transferred by several mechanisms through a structure of mean cross-sectional area A_m, conductance equals the gross rate of heat transfer divided by the temperature drop between its faces.

Thermal conductivity: Measure of the ability of a material to conduct heat; measured in flow of Btus per hour through a unit cross section or unit thickness with 1° F of temperature difference across this thickness. For refractory and insulation materials, typical units are Btu-in./ft²-hr-°F. Other acceptable units are Btu-ft²-°F.

Thermal expansion: An increase in volume of linear dimensions caused by heating the material.

Thermal shock: Denotes a sudden temperature change.

Thermoelasticity: Rubberlike elasticity that a rigid plastic displays; it is caused by elevated temperatures.

Thermoplastic (n): A plastic that repeatedly can be softened by heating and hardened by cooling through a temperature range characteristic of the plastic, and that in the softened state can be shaped by flow into articles by molding or extrusion.

Thermoplastic (adj.): Capable of being repeatedly softened by heating and hardened by cooling through a temperature range characteristic of the plastic, and that in the softened state can be shaped by flow into articles by molding or extrusion. (*Note:* Thermoplastic applies to those materials whose change upon heating is substantially physical.)

Thermoset (n): A plastic that, when cured by application of heat or by chemical means, changes into a substantially infusible and insoluble product.

Thermoset (adj.): Pertaining to the state of a resin in which it is relatively infusible.

Thermosetting: Capable of being changed into a substantially infusible or insoluble product when cured by application of heat or by chemical means.

Tolerance: The total range of variation permitted; the upper and lower limits between which a dimension must be maintained.

Tracer yarn: Strand of glass fiber colored differently from the remainder of the roving package. It is a means of determining whether equipment used to chop and spray glass fibers are functioning properly and provides a check on quality and thickness control.

Trap: A fitting or device designed and constructed to provide, when properly vented, a liquid seal that will prevent the back passage of air without materially affecting the flow of sewage or waste water through it.

Tubing (plastic): A particular size of plastic pipe in which the outside diameter is essentially the same as that of copper tubing.

Urethane plastics: Group of plastics composed of resins derived from the condensation of organic isocyanates with compounds containing hydroxyl groups.

Vacuum: Any pressure less than that exerted by the atmosphere; it may be termed a negative pressure.

Vacuum forming: Fabrication process in which plastic sheets are transformed to desired shapes by inducing flow; accomplished by reducing the air pressure on one side of the sheet.

Valve: A device that regulates the flow of fluids through piping by opening, closing, or obstructing ports or passageways.

Valve positioner: Auxiliary servo device that allows precision positioning of a control valve stem. It is used in conjunction with a standard valve operator (e.g., a diaphragm motor). Its purpose is to overcome stuffing box friction and stem thrust caused by fluid pressure.

Van Stone flange: A fitting flange whose drilled back plate turns 360° in order to facilitate the joining of one flange to another flange.

Velocity: Time rate of motion in a given direction.

Velocity head: Velocity pressure expressed in feet of column of the flowing fluid.

Vinyl chloride plastics: Plastics based on resins made by the polymerization of vinyl chloride and copolymerization of vinyl chloride with other unsaturated compounds, the vinyl chloride being in greatest amount by weight.

Virgin material: A plastic material in the form of pellets, granules, powder, floc, or liquid that has not been subjected to use or processing other than that required for its original manufacture.

Viscosity: A measure of the tendency of a fluid to resist shear. The unit of viscosity is the poise, which is defined as the resistance (in dynes per square centimeter of its surface) of one layer of fluid to the motion of a parallel layer one centimeter away and with a relative velocity of one centimeter per second.

Water absorption: The percentage of water absorbed by an immersed specimen in a given time.

Waterhammer: The forces, pounding noises, and vibration that develop in a piping system when a column of incompressible liquid flowing through a pipe line at a given pressure and velocity is stopped abruptly.

Waterhammer arrester: A device, other than an air chamber, designed to provide protection against excessive surge pressure.

Weathering: Exposure of a plastic to outdoor conditions.

Weld- or knit-line: A mark on a molded plastic formed by the union of two or more streams of plastic flowing together.

Yield value: Also called yield stress; force necessary to initiate flow in a plastic.

Young's modulus of elasticity: The modulus of elasticity in tension. The ratio of stress in a material subjected to deformation.

B

Appendix

Abbreviations

A. Plastic Material Abbreviations

ABS = acrylonitrile–butadiene–styrene
ADC = allyl diglycol carbonate
ASA = acrylonitrile–styrene–acrylic
CA = cellulose acetate
*CAB = cellulose acetate butyrate
CAP = cellulose acetate propionate
CN = cellulose nitrate
CP = cellulose propitionate
CPET = crystallized polyethylene terephthalate
*CPVC = chlorinated polyvinyl chloride
CTFE = chlorotrifluorethylene
DAP = diallyl phthalate
EC = ethyl cellulose
ECTF = ethylene–chlorotrifluoroethylene
EEA = ethylene–ethyl acrylate
EMA = ethylene–methyl acrylate
EPDM = ethylene propylene diene monomer
EPS = expandable polystyrene
ETFE = ethylenetetrafluoroethylene
EVA = ethylene–vinyl acetate
*FEP = fluorinated ethylenepropylene
*FRP = fiberglass-reinforced plastics
*HDPE = high-density polyethylene
HIPS = high-impact polystyrene
*HMW–HDPE = high-molecular-weight high-density polyethylene
LCP = liquid crystal polymers
*LDPE = low-density polyethylene
LLDPE = linear low-density polyethylene
MBS = methacrylate–butadiene–styrene
MDI = methylene diisocyanate
*MDPE = medium-density polyethylene

MMA = methyl methacrylate
MPPO = modified polyethylene oxide
OPET = oriented polyethylene terephthalate
OPP = oriented polypropylene
OSA = olefin-modified styrene–acrylonitrite
PA = polyamide
PAI = polyamideimide
PAN = polyacrylonitrile
*PB = polybutylene
PBT = polybutylene terephthalate
PC = polycarbonate
PCTFE = polychlorotrifluoroethylene
PFA = perfluoroalkoxy
*PE = polyethylene
PEEK = polyetheretherketone
PEO = biaxially oriented polyethylene
PES = polyethersulfone
PET = polyethylene terephthalate
PEX = cross-linked polyethylene
PMMA = polymethyl methacrylate
*PP = polypropylene
*PPFR = polypropylene flame retardant
PPS = polyphenylene sulfide
PS = polystyrene
PSO = polysulfone
*PTFE = polytetrafluoroethylene
PTMT = polytetramethylene terephthalate
PUR = polyurethane
PVAC = polyvinyl acetate
PVAL = polyvinyl alcohol
PVB = polyvinyl butyral
*PVC = polyvinyl chloride
*PVDC = polyvinylidene chloride
*PVDF = polyvinylidene fluoride

---

*Plastic piping materials.

*Plastic piping materials.

PVF = polyvinyl fluoride
RTRP = reinforced thermosetting resin pipe
SAN = styrene–acrylonitrile
SBR = styrene–butadiene rubber
SMA = styrene–maleic anhydride
***SR** = styrene rubber
TFE = tetrafluoroethylene
TPE = thermoplastic elastomers
TPO = thermoplastic olefin
TPU = thermoplastic polyurethanes
***UHMWPE** = ultrahigh-molecular-weight
                    polyethylene
UHMW-HDPE = ultrahigh-molecular weight high-
                    density polyethylene
VCM = vinyl chloride monomer
VDC = vinylidene chloride

B. List of Piping and Engineering Terms

abs = absolute
Amer Std = American Standard
avg = average
AWD = acid waste drainage
BE = belled end
bbl = barrel
Btu = British thermal unit
bush = bushing
C = coefficient constant
°C = Celsius or centigrade (degree)
C to F = center to face
cfm = cubic feet per minute
cfs = cubic feet per second
cl = close (as in close nipple)
CI = cast iron
cm = centimeter(s)
cm² = square centimeter(s)
cm³ = cubic centimeter(s)
conc = concentric
coupl = coupling
CPI = Chemical Process Industry
cts = copper tube size
cu ft = cubic feet
cu in. = cubic inch(es)
deg = degree
diam = diameter
dwg = drawing
DWV = drain, waste, vent
eccen = eccentric
ell = elbow
°F = fahrenheit [degree(s)]

---

*Plastic piping materials.

fig = figure
FIPT = female iron pipe thread
flg = flange
flgd = flanged
ft = feet
ft² = square feet
ft³ = cubic feet
g = gauge (or gage)
gal = gallon
gph = gallon(s) per hour
gpm = gallon(s) per minute
HDS = hydrostatic design stress
hex = hexagonal
Hg = mercury
hr = hour
ID = inside diameter
in. = inch(es)
inc = increaser
IPS = iron pipe size
kw = killowatt(s)
lb = pound(s)
max = maximum
min = minimum
min = minute
mfr = manufacturer
MIPT = male iron pipe thread
mpt = male pipe thread
mtd = mounted
NA = not applicable
NPS = nominal pipe size
NR = not recommended
OD = outside diameter
OWG = oil, water, gas
psi = pounds per square inch
psig = pounds per square inch, gauge
PE = plain end
R = recommended
rad = radius
s = socket
red = reducing
scd = screwed
sch = schedule
sched = schedule
sec = second
slip = female socket
soc = socket
spec = specification
spg = spigot
sq = square
sq ft = square feet
SSU = seconds Saybolt universal
std = standard
t = thread

TBE = threaded both ends
thd = threaded
TOE = threaded one end
trans = transmission
vel = velocity
vlv = valve
WOG = water, oil, gas
wt = weight
WWP = working water pressure

C. Piping Standards/Code Agencies

AASHTO = American Association of State Highways and Transportation Officials
AGA = American Gas Association
ANSI = American National Standards Institute, Inc.
API = American Petroleum Institute
ASAE = American Society of Agricultural Engineers
ASHRAE = American Society of Heating, Refrigeration and Air Conditioning Engineers
ASHVE = American Society of Heating and Ventilating Engineers
ASME = American Society of Mechanical Engineers
ASNP = American Standard National Plumbing
ASPE = American Society of Plumbing Engineers
ASSE = American Society of Sanitary Engineers
ASTM = American Society for Testing and Materials
AWWA = American Water Works Association
BOCA = Building Officials and Code Administrators International, Inc.
BS = British Standards Institution
CGSB = Canadian Government Specifications Board
CS = Commercial Standard (superseded by term *Product Standard*)
CSA = Canadian Standards Association
DIN = Deutsches Institut für Normung (German Industrial Norms)
DOD = Department of Defense
DOT–OPSO = Department of Transportation–Office of Pipeline Safety Operations

FHA = Federal Housing Administration
FM = Factory Mutual
FS = Federal Standards
HUD = Department of Housing and Urban Development
IAPMO = International Association of Plumbing and Mechanical Officials
ICPO = International Conference of Building Officials
ISO = International Standards Organization
JIS = Japanese Industrial Standards
MIL = Department of Defense–Military Standard
MSS = Manufacturers Standardization Society of the Valve and Fittings Industry
NAPD = National Association of Plastic Distributors
NBS = National Bureau of Standards
NEMA = National Electrical Manufacturers Association
NFPA = National Fire Protection Association
NSF = National Sanitary Foundation
NSPI = National Swimming Pool Institute
PDI = Plumbing and Drainage Institute
PHCC = National Association of Plumbing–Heating–Cooling Contractors
PPFA = Plastic Pipe and Fittings Association
PPI = Plastic Piping Institute
PS = Product Standard
SBCC = Southern Building Code Congress
SCS = Soil Conservation Service of U.S. Department of Agriculture
SIA = Sprinkler Irrigation Association
SPE = The Society of Plastic Engineers
SPI = The Society of the Plastics Industry, Inc.
UL = Underwriters Laboratories, Inc.
UNI-Bell = Uni-Bell PVC Pipe Association
USASI = United States of America Standards Institute
VMA = Valve Manufacturers Association of America
VPS = Voluntary Product Standards
WUC = Western Underground Committee

C

Appendix

Commonly Used Plastics Trade Names

Trade Name	Materials	Manufacturer
Alathon	polyethylene resins	Dupont
Atlac	fiberglass-reinforced resins	ICI Americas
Bakelite	plastic resins	Union Carbide
Boltaron	vinyl resins	General Tire
Cab-XL	cross-linked polyethylene	Cabot Corp.
Celcon	acetal copolymers	Celanese
Cyclolac	ABS plastics	Marbon Chemical
Dacron	polyester textile fibers	Dupont
Delrin	acetal homopolymer	Dupont
Dion	fiberglass-reinforced resins	Koppers Co.
Fluon	fluorocarbon resins	ICI Americas
Fluorel	fluorocarbon elastomers	3M Company
Formica	melamine laminates	Formica Corp.
Freon	gaseous or liquid fluorocarbons	Dupont
Furan	fiberglass-reinforced resins	Quaker Oats
Fuseal	polypropylene acid-waste drainage systems	R. & G. Sloane
Geon	vinyl resins	B. F. Goodrich
Halar	fluorocarbon resins (PCTFE)	Allied Corp.
Halon	fluorocarbon resins (PTFE)	Allied Corp.
Haveg	fiberglass-reinforced resins	Ametek
Hetron	fiberglass-reinforced resins	Ashland Chemical Co
Hi Temp	chlorinated polyvinylchloride	B. F. Goodrich
Hypalon	chlorosulfonated polyethylene	Dupont
Kel-F	fluorocarbon resins (PCTFE)	3M Company
Kodel	polyester textile fiber	Eastman Chem.
Kraylastic	ABS resins	USS Chemical
Kydex	acrylic-modified PVC resins	Rohm & Haas
Kynar	polyvinylidene fluoride resins	Pennwalt

Trade Name	Materials	Manufacturer
Lab-Line	polypropylene acid-waste drainage system	Enfield Indust.
Lexan	polycarbonate resins	General Electric
Lucite	acrylic resins	Dupont
Lustran	ABS resins	Monsanto
Marlex	polyethylene resins	Phillips Chemical
Micarta	phenolic compounds	Allied Plastics
Mylar	polyester resins	Dupont
Noryl	modified phenylene oxides	General Electric
Nylatron	polyamide resins	Polymer Corp.
Nylon	polyamide resins	Dupont
Orlon	acrylic textile fibers	Dupont
PEEK	polyetheretherketone	ICI
Penton	chlorinated polyether	Hercules Inc.
Plexiglass	acrylic resins	Rohm & Haas
Quacorr	fiberglass-reinforced resins	Quaker Oats
Rexene	polyethylene compound	Rexene Co.
Ryton	polyphenylene sulfide resin	Phillips Chemical
Saran	polyvinylidene chloride resin	Dow Chemical
Tenite	plastic resins	Eastman Chemical
Teflon	fluorocarbon resins (TFE, PFA, FEP)	Dupont
Tefzel	fluorocarbon resin (ETFE)	Dupont
Tygon	vinyl tubing	Norton Co.
Ultem	thermoplastic polyetherimide	General Electric
Valox	plastic resins	General Electric
Vectra	liquid crystal aromatic polyester	Celanese
Viton	fluorocarbon elastomers	Dupont
Vulcathene	polyethylene drainage system	Vulcathene Ltd.
Xydar	liquid crystal aromatic polyester	Dartco
Zytel	polyamide resins	Dupont

D
Appendix

CHEMICAL RESISTANCE OF COMMONLY USED PLASTIC PIPING MATERIALS

| | IN DEGREES FAHRENHEIT | | | | | | | | | | | | | | | | | | | | | | | | |
Chemical	PVC 70	PVC 140	CPVC 70	CPVC 140	CPVC 185	Polypro-pylene 70	150	180	PVDF 70	150	250	Poly-eth-ylene 70	140	Fiberglass-Reinforced Epoxy 70	150	250	Fiberglass-Reinforced Vinyl/Polyester 70	150	250	Neoprene 70	185	EPDM 70	185	Viton® 70	185
Acetaldehyde	NR	NR	·	·	·	R	·	·	R	R	·	R	NR	NR	NR	·	·	·	·	NR	NR	R	R	NR	NR
Acetamide	·	·	R	·	·	·			·	·	·	·	·	·	·	·	·	·	·	R	NR	R	R	R	NR
Acetate Solvents	NR	NR	NR	NR	NR	NR	NR	NR	R	·	NR	·	·	·	·	·	·	·	·	·	·	·	·	·	·
Acetic Acid, 10%	R	R	R	R	NR	R	R	R	R	R	R	R	R	R	R	·	R	R	R	NR	NR	R	R	NR	NR
Acetic Acid, 20%	R	R	R	R	NR	R	R	R	R	R	R	R	R	R	R	·	R	R	R	R	R	R	R	NR	NR
Acetic Acid, 50%	R	R	·	·	NR	R	R	R	R	R	R	R	R	R	·	·	R	R	·	R	NR	R	R	NR	NR
Acetic Acid, 80%	R	R	·	·	NR	R	·	·	R	R	NR	R	R	R	NR	NR	R	R	·	R	NR	R	R	NR	NR
Acetic Acid, Glacial	NR	NR	NR	NR	NR	R	NR	NR	R	R	NR	R	R	R	R	·	R	·	·	NR	NR	R	NR	NR	NR
Acetic Anhydride	NR	NR	NR	NR	NR	R	·	·	R	NR	NR	R	R	NR	NR	NR	NR	NR	NR	NR	NR	NR	NR	NR	NR
Acetone	NR	NR	NR	NR	NR	R	NR	NR	NR	NR	NR	NR	NR	R	R	NR	NR	NR	NR	NR	NR	R	R	NR	NR
Acetophenone	·	·	·	·	·	R			R	NR	NR	·	·	·	·	·	·	·	·	NR	NR	R	R	NR	NR
Acetyl Chloride	NR	NR	NR	NR	NR	·	·	·	R	R	·	·	·	·	·	·	·	·	·	NR	NR	NR	NR	R	R
Acetylene	R	R	R	R	·	R	·	·	R	·	·	NR	NR	·	·	·	·	·	·	NR	NR	R	R	R	R
Acrylic Acid Ethyl Ester	NR	NR	NR	NR	NR	·	·	·	·	·	·	·	·	R	·	·	·	·	·	·	·	·	·	·	·
Acrylonitrile	·	·	·	·	·	·	·	·	R	NR	NR	·	·	·	·	·	·	·	·	NR	NR	NR	NR	NR	NR
Adipic 105 Acid	R	R	R	R	R	R	R	·	R	·	·	R	·	R	R	·	R	R	·	R	R	R	R	·	·
Alcohol Allyl	NR	NR	NR	NR	NR	R	·	·	R	R	·	R	·	·	·	·	·	·	·	·	·	R	NR	R	R
Alcohol Amyl	NR	NR	R	·	NR	R	·	·	R	R	R	R	R	R	R	·	R	R	R	NR	NR	R	R	NR	NR
Alcohol Benzyl	·	·	·	·	·	R	R	·	R	R	R	·	·	NR	NR	NR	R	·	·	NR	NR	R	R	R	R
Alcohol, Butyl, Primary	R	·	R	R	·	R	R	R	R	R	R	R	R	R	R	·	R	R	·	R	R	R	R	R	R
Alcohol, Butyl, Secondary	R	·	R	R	NR	R	R	R	R	R	R	R	R	R	R	·	R	R	·	R	R	R	R	R	R
Alcohol, Diacetone	·	·	·	·	·	R	NR	NR	R	NR	NR	NR	NR	·	·	·	·	·	·	R	NR	R	R	NR	NR
Alcohol, Ethyl	R	R	R	R	R	R	R	R	R	R	R	NR	NR	R	R	·	R	R	·	R	R	R	R	NR	NR
Alcohol, Hexyl	R	R	·	·		·	·	·	·	·	·	·	·	·	·	·	·	·	·	R	NR	NR	NR	R	R
Alcohol, Isopropyl	R	R	·	·		R	·	·	·	·	·	R	R	R	R	·	R	R	·	R	NR	R	R	R	R
Alcohol, Methyl	R	R	R	R	·	R	R	R	R	R	R	R	NR	R	R	·	R	·	·	R	R	R	R	NR	NR
Alcohol, Propargyl	R	R	·	·	·	NR	NR	NR	·	·	·	R	R	·	·	·	·	·	·	·	·	·	·	·	·
Alcohol, Propyl	R	·	R	·	NR	R	·	NR	R	NR	NR	·	·	·	·	·	·	·	·	R	R	R	R	R	R
Allyl Chloride	NR	NR	NR	NR	NR	R	·	·	R	R	·	R	R	·	·	·	·	·	·	·	·	R	·	·	·
Alum	R	R	R	R	R	R	R	R	R	R	R	R	R	R	R	R	R	R	·	R	R	R	R	·	·
Alum, Ammonium	NR	NR	NR	NR	NR	R	R	R	R	R	R	R	R	R	R	·	R	R	·	·	·	R	R	R	R
Alum, Chrome	R	R	R	R	R	R	R	R	R	R	R	R	R	R	R	·	R	R	·	·	·	R	R	R	R
Alum, Potassium	R	R	R	R	R	R	R	R	R	R	R	R	R	R	R	·	R	R	·	·	·	R	R	R	R
Aluminum Chloride	R	R	R	R	R	R	R	R	R	R	R	R	R	R	R	·	R	R	·	R	R	R	R	R	R
Aluminum Fluoride	R	R	R	R	R	R	R	R	R	R	R	R	R	R	R	·	R	·	·	R	R	R	R	R	R
Aluminum Hydroxide	R	R	R	R	R	R	R	R	R	R	R	R	R	R	R	·	R	·	·	·	·	R	R	R	R
Aluminum Nitrate	R	R	R	R	R	R	R	R	R	R	R	R	R	R	R	·	R	R	·	R	R	R	R	R	R
Aluminum Oxychloride	R	R	R	R	R	R	R	R	·	·	·	R	R	·	·	·	·	·	·	·	·	·	·	NR	NR

Notes—Designations used are: R = Recommended; NR = Not Recommended; · = No data. All temperatures are in °F, and all chemicals are in solution in the pure or concentrated state unless indicated otherwise.

Chemical	PVC 70	PVC 140	CPVC 70	CPVC 140	CPVC 185	Polypropylene 70	Polypropylene 150	Polypropylene 180	PVDF 70	PVDF 150	PVDF 250	Polyethylene 70	Polyethylene 140	FRP Epoxy 70	FRP Epoxy 150	FRP Epoxy 250	FRP Vinyl/Polyester 70	FRP Vinyl/Polyester 150	FRP Vinyl/Polyester 250	Neoprene 70	Neoprene 185	EPDM 70	EPDM 185	Viton® 70	Viton® 185
Aluminum Sulfate	R	R	R	R	R	R	R	R	R	R	R	R	R	R	R	R	R	R	·	R	R	R	R	R	R
Aluminum Dichromate	R	·	·	·	·	·	·	·	·	·	·	·	·	·	·	·	·	·	·	·	·	·	·	·	·
Aluminum Fluoride, 10%	R	R	·	·	·	·	·	·	·	·	·	R	R	R	R	·	R	·	·	·	·	·	·	·	·
Aluminum Fluoride, 25%	R	R	·	·	·	R	R	·	·	·	·	·	·	R	R	·	R	·	·	R	·	·	·	·	·
Aluminum Hydroxide, 10%	R	R	R	R	R	R	R	R	·	·	·	·	·	R	R	·	R	R	·	R	·	·	·	·	·
Aluminum Hydroxide, 25%	R	R	R	R	R	R	R	R	·	·	·	·	·	R	R	·	R	R	·	R	·	·	·	·	·
Aluminum Metaphosphate	R	R	R	R	R	R	R	·	·	·	·	·	·	R	R	·	·	·	·	·	·	·	·	·	·
Aluminum Nitrate	R	R	R	R	R	R	R	R	R	R	R	R	R	R	R	R	R	R	·	R	R	R	R	R	R
Aluminum Persulfate	R	R	R	R	R	R	R	·	·	·	·	·	·	R	R	·	R	·	·	·	·	·	·	·	·
Aluminum Phosphate	R	R	R	R	R	R	R	R	·	·	·	·	·	R	R	·	R	·	·	R	R	R	R	R	R
Aluminum Sulfate	R	R	R	R	R	R	R	R	·	·	·	·	·	R	R	·	R	R	·	R	R	R	R	R	R
Aluminum Sulfide	R	R	R	R	R	R	·	·	·	·	·	·	·	R	R	·	·	·	·	·	·	·	·	·	·
Aluminum Thiocyanate	R	R	R	R	R	R	·	·	R	NR	NR	R	R	R	R	·	R	·	·	·	·	·	·	·	·
Ammonia Gas, Dry	R	R	R	R	R	R	R	·	R	R	R	R	R	R	R	·	R	·	·	R	R	R	R	NR	NR
Ammonia, Aqua, 10%	R	R	R	R	R	R	R	·	R	R	R	·	·	R	R	·	R	·	·	R	NR	R	R	NR	NR
Ammonia, Liquid	NR	NR	NR	NR	NR	R	R	·	R	R	R	·	·	·	·	·	·	·	·	R	NR	R	R	NR	NR
Ammonium Acetate	R	R	R	R	·	R	R	·	R	R	·	·	·	·	·	·	·	·	·	·	·	R	R	NR	NR
Ammonium Bifluoride	R	R	R	R	R	R	R	R	R	R	R	R	R	·	·	·	·	·	·	·	·	·	·	·	·
Ammonium Bisulfide	R	R	·	·	·	·	·	·	·	·	·	·	·	·	·	·	·	·	·	·	·	·	·	·	·
Ammonium Carbonate	R	R	R	R	R	R	R	R	R	R	R	R	R	R	R	·	R	·	·	R	R	R	R	·	·
Ammonium Chloride	R	R	R	R	R	R	R	·	R	R	R	R	R	R	R	·	R	R	·	R	R	R	R	R	R
Ammonium Fluoride, 10%	R	R	R	R	R	R	R	R	R	R	R	R	R	NR	NR	NR	R	R	·	·	·	R	R	·	·
Ammonium Hydroxide	R	R	R	R	R	R	R	R	R	R	R	R	R	R	R	·	R	R	·	R	R	R	R	R	·
Ammonium Nitrate	R	R	R	R	R	R	R	R	R	R	R	R	R	R	R	R	R	R	R	R	R	R	R	NR	NR
Ammonium Phosphate	R	R	R	R	R	R	R	R	R	R	·	·	·	R	R	·	R	R	R	R	R	R	R	R	R
Ammonium Sulfate	R	R	R	R	R	R	R	R	R	R	R	R	R	R	R	R	R	R	·	R	R	R	R	NR	NR
Ammonium Sulfide	R	R	R	R	R	R	R	R	R	R	R	R	R	NR	NR	NR	NR	NR	NR	R	R	R	R	NR	NR
Ammonium Thiocynate	R	R	R	R	R	R	R	·	·	·	·	·	·	R	R	·	·	·	·	·	·	·	·	R	R
Amyl Acetate	NR	NR	NR	NR	NR	NR	NR	NR	R	NR	NR	R	NR	R	·	·	·	·	·	NR	NR	NR	NR	NR	NR
Amyl Chloride	NR	NR	NR	NR	NR	NR	NR	NR	R	R	R	NR	NR	·	·	·	NR	NR	NR	·	·	NR	NR	R	R
Aniline	NR	NR	NR	NR	NR	R	R	·	R	NR	NR	R	NR	R	·	·	·	·	·	NR	NR	R	R	NR	NR
Aniline Chlorohydrate	NR	NR	·	·	·	·	·	·	·	·	·	·	·	·	·	·	·	·	·	·	·	·	·	·	·
Aniline Hydrochloride	NR	NR	NR	NR	NR	R	·	·	R	·	·	·	·	R	R	·	NR	NR	NR	R	R	NR	NR	NR	NR
Anthraquinone	R	R	·	·	·	NR	NR	NR	·	·	·	·	·	·	·	·	·	·	·	·	·	·	·	R	R
Anthraquinone Sulfonic Acid	R	R	R	R	R	R	NR	NR	R	·	·	·	·	·	·	·	·	·	·	·	·	·	·	R	R
Antimony Trichloride	R	NR	NR	NR	NR	R	·	·	R	NR	NR	R	R	R	R	·	R	·	·	·	·	R	·	R	R
Aqua Regia	NR	NR	NR	NR	NR	NR	NR	NR	R	·	·	NR	NR	NR	NR	NR	·	·	·	NR	NR	NR	NR	R	R
Arsenic Acid	R	R	R	R	R	R	·	·	R	R	R	R	·	·	·	·	·	·	·	R	R	R	R	R	R
Aryl Sulfonic Acid	NR	NR	NR	·	·	NR	·	·	NR	·	·	·	·	·	·	·	·	·	·	·	·	·	·	R	R
Barium Carbonate	R	R	R	R	R	R	R	R	R	R	R	R	R	R	R	·	R	R	·	·	·	R	R	·	·
Barium Chloride	R	R	R	R	R	R	R	R	R	R	R	R	R	R	R	·	R	R	·	R	R	R	R	R	R
Barium Hydroxide, 10%	R	R	R	R	R	R	R	R	R	R	R	R	R	R	R	·	·	·	·	R	R	R	R	R	R
Barium Hydroxide, Conc.	R	R	R	R	R	R	R	R	R	R	R	R	R	R	·	·	NR	NR	NR	·	·	·	·	·	·
Barium Nitrate	R	·	·	·	·	·	·	·	·	·	·	·	·	·	·	·	·	·	·	·	·	R	R	R	R
Barium Sulfate	R	R	R	R	R	R	R	R	R	R	R	R	R	R	R	R	R	R	·	R	R	R	R	R	R
Barium Sulfide	R	R	R	R	R	R	R	R	R	R	R	R	R	R	R	R	R	R	·	R	R	R	R	R	R
Beer	R	R	R	R	R	R	R	R	R	R	R	R	R	R	R	·	R	·	·	R	R	R	R	R	R
Beet Sugar Liquors	R	R	R	R	R	R	R	R	R	R	R	R	R	R	·	·	·	·	·	R	NR	R	R	R	R
Benzaldehyde, 10%	R	R	·	·	·	R	·	·	R	·	·	·	·	R	R	·	NR	NR	NR	NR	NR	R	R	NR	NR
Benzaldehyde, above 10%	NR	NR	NR	NR	NR	R	R	·	R	NR	NR	R	R	NR	NR	NR	NR	NR	NR	·	·	·	·	·	·
Benzalkonium Chloride	R	·	·	·	·	·	·	·	·	·	·	·	·	·	·	·	·	·	·	·	·	·	·	·	·
Benzene, Benzol	NR	NR	NR	NR	NR	R	NR	NR	R	NR	NR	NR	NR	R	R	·	·	·	·	NR	NR	NR	NR	R	R
Benzene Sulfonic Acid, 10%	R	R	R	R	·	R	R	R	R	NR	NR	R	R	R	·	·	R	R	·	R	NR	NR	NR	R	R
Benzoic Acid	R	R	R	R	R	R	R	·	R	R	R	R	R	R	·	·	R	R	·	NR	NR	NR	NR	R	R
Benzyl Alcohol	·	·	·	·	·	R	·	·	R	R	·	·	·	NR	NR	NR	NR	NR	NR	R	NR	R	R	R	R
Bismuth Carbonate	R	R	R	R	R	R	R	·	R	R	R	R	R	·	·	·	·	·	·	·	·	·	·	·	·
Black Liquor	R	R	R	R	R	R	R	NR	R	R	R	R	R	R	R	·	·	·	·	·	·	NR	NR	R	R
Bleach, 12.5%, Active Cl_2	R	R	R	R	R	R	NR	NR	R	R	R	R	R	R	R	·	·	·	·	NR	NR	R	R	R	R
Borax	R	R	R	R	R	R	·	·	R	R	R	R	R	R	R	·	·	·	·	R	R	R	R	R	R
Boric Acid	R	R	R	R	R	R	R	R	R	R	R	R	R	R	R	·	R	·	·	R	R	R	R	R	R
Brine Acid	R	R	R	R	R	R	R	·	R	R	R	R	R	·	·	·	·	·	·	R	R	R	R	R	R
Bromic Acid	R	R	R	R	R	NR	NR	NR	R	R	·	R	R	R	R	·	·	·	·	R	R	R	R	R	R
Bromine Liquid	NR	NR	NR	NR	NR	NR	NR	NR	R	R	·	NR	NR	·	·	·	·	·	·	NR	NR	NR	NR	R	R

Notes—Designations used are: R = Recommended; NR = Not Recommended; · = No data. All temperatures are in °F, and all chemicals are in solution in the pure or concentrated state unless indicated otherwise.

IN DEGREES FAHRENHEIT

Chemical	PVC		CPVC			Polypropylene			PVDF			Polyethylene		Fiberglass-Reinforced Epoxy			Fiberglass-Reinforced Vinyl/Polyester			Neoprene		EPDM		Viton®	
	70	140	70	140	185	70	150	180	70	150	250	70	140	70	150	250	70	150	250	70	185	70	185	70	185
Bromine Vapor, 25%	NR	NR	NR	NR	NR	NR	NR	NR	R	R	·	R	R	·	·	·	·	·	·	·	·	·	·	·	·
Bromine, Water	NR	NR	NR	NR	NR	NR	NR	NR	R	R	·	NR	NR	·	·	·	·	·	·	NR	NR	NR	NR	R	R
Bromobenzene	NR	NR	·	·	·	·	·	·	·	·	·	·	·	·	·	·	·	·	·	NR	NR	NR	NR	R	R
Bromotoluene	NR	NR	·	·	·	NR	NR	NR	·	·	·	·	·	·	·	·	·	·	·	·	·	·	·	·	·
Butadiene	NR	NR	R	R	·	R	R	·	R	R	·	·	·	R	·	·	·	·	·	NR	NR	NR	NR	R	R
Butane	R	NR	R	·	NR	R	R	·	·	·	·	NR	NR	R	R	·	·	·	·	R	R	NR	NR	R	R
Butanols	R	R	R	R	R	R	R	R	R	R	R	·	·	·	·	·	·	·	·	·	·	·	·	·	·
Butyl Acetate	R	NR	NR	NR	NR	NR	NR	NR	R	NR	NR	NR	NR	R	·	·	·	·	·	NR	NR	NR	NR	NR	NR
Butyl Alcohol	NR	NR	R	NR	NR	R	R	R	R	R	R	R	R	R	·	·	·	·	·	R	R	R	NR	R	R
Butyl Cellosolve	R	·	·	·	·	·	·	·	·	·	·	·	·	R	·	·	R	·	·	NR	NR	R	R	NR	NR
Butyl Phthalate	NR	NR	NR	NR	NR	R	R	R	R	R	R	·	·	·	·	·	·	·	·	·	·	·	·	·	·
Butylene	R	NR	R	NR	NR	NR	NR	NR	R	R	R	·	·	·	·	·	·	·	·	NR	NR	NR	NR	R	R
Butyl Phenol	NR	NR	R	NR	NR	R	·	·	R	R	·	·	·	·	·	·	·	·	·	·	·	·	·	·	·
Butyl Stearate	R	·	·	·	·	·	·	·	·	·	·	·	·	·	·	·	·	·	·	NR	NR	NR	NR	R	R
Butyne Diol	NR	NR	R	NR	NR	R	R	·	R	·	·	·	·	R	·	·	·	·	·	·	·	·	·	·	·
Butyric Acid, 25%	NR	NR	R	NR	NR	R	R	R	R	R	R	NR	NR	R	R	·	R	R	·	·	·	R	NR	R	NR
Cadmium Cyanide	R	R	R	R	R	·	·	·	·	·	·	·	·	R	·	·	·	·	·	·	·	·	·	·	·
Caffeine Citrate	R	·	·	·	·	·	·	·	·	·	·	·	·	R	·	·	·	·	·	·	·	·	·	·	·
Calcium Bisulfide	R	R	R	R	R	R	R	R	R	R	R	R	R	R	R	·	·	·	·	·	·	·	·	·	·
Calcium Bisulfite	R	R	R	R	R	R	R	R	R	R	R	R	R	R	R	·	R	R	·	R	R	NR	NR	R	R
Calcium Carbonate	R	R	R	R	R	R	R	R	R	R	R	R	R	R	R	·	R	·	·	R	R	R	R	R	R
Calcium Chlorate	R	R	R	R	R	R	R	R	R	R	R	R	R	R	R	·	R	R	·	·	·	·	·	R	R
Calcium Chloride	R	R	R	R	R	R	R	R	R	R	R	R	R	R	R	R	R	R	R	R	R	R	R	R	R
Calcium Hydroxide, 50%	R	R	R	R	R	R	R	R	R	R	R	R	R	R	·	·	R	R	·	R	R	R	R	R	R
Calcium Hypochlorite	R	·	R	R	R	R	·	NR	R	R	R	R	R	NR	NR	·	R	R	·	NR	NR	R	R	R	R
Calcium Hydroxide, Conc.	R	R	R	R	R	R	R	R	R	R	R	R	R	R	·	·	·	·	·	·	·	·	·	·	·
Calcium Nitrate	R	R	R	R	R	R	R	R	R	R	R	R	R	R	R	R	R	R	R	R	R	R	R	R	R
Calcium Oxide	R	R	R	R	·	·	·	·	·	·	·	R	R	·	·	·	·	·	·	·	·	R	R	·	·
Calcium Sulfate	R	R	R	R	R	R	R	R	R	R	R	R	R	R	R	R	R	R	R	·	·	R	R	·	·
Camphor Crystals	R	·	·	·	·	·	·	·	·	·	·	NR	NR	·	·	·	·	·	·	·	·	·	·	·	·
Cane Sugar Liquors	R	R	R	R	R	R	R	R	R	R	R	R	R	·	·	·	·	·	·	R	R	R	R	R	R
Carbitol	R	·	·	·	·	·	·	·	·	·	·	·	·	·	·	·	·	·	·	R	NR	R	NR	R	NR
Caprylic Acid	·	·	·	·	·	·	·	·	R	R	NR	·	·	NR	NR	NR	·	·	·	·	·	·	·	·	·
Carbon Bisulfide	NR	NR	NR	NR	NR	NR	NR	NR	R	NR	NR	NR	NR	NR	NR	NR	·	·	·	NR	NR	NR	NR	R	R
Carbon Dioxide, Dry	R	R	R	R	R	R	R	R	R	R	R	R	R	R	R	·	·	·	·	R	NR	R	NR	R	R
Carbon Dioxide, Wet	R	R	R	R	R	R	R	R	R	R	R	R	R	R	R	·	·	·	·	·	·	·	·	·	·
Carbon Disulfide	NR	NR	NR	NR	NR	NR	NR	NR	R	NR	NR	NR	NR	R	·	NR	R	·	·	·	·	NR	NR	R	R
Carbon Monoxide	R	R	R	R	R	R	R	·	R	R	R	R	R	·	·	·	R	R	·	R	NR	R	R	R	R
Carbon Tetrachloride	NR	NR	NR	NR	NR	NR	NR	NR	R	NR	NR	NR	NR	R	·	·	R	·	·	NR	NR	NR	NR	R	R
Carbonic Acid	R	R	R	R	R	R	R	R	·	·	·	R	R	R	R	·	·	·	·	R	R	R	R	R	R
Castor Oil	R	R	R	R	R	R	·	NR	R	R	R	R	R	R	R	·	R	R	·	R	R	R	NR	R	R
Caustic Potash	R	R	R	R	R	R	R	R	R	R	R	·	·	R	·	·	R	R	·	·	·	·	·	NR	NR
Caustic Soda	R	R	R	R	R	R	R	R	R	R	·	R	·	R	R	·	NR	NR	NR	·	·	·	·	·	·
Cellosolve	R	NR	R	R	·	R	·	·	R	R	R	R	·	·	·	·	·	·	·	NR	NR	R	NR	NR	NR
Cellosolve Acetate	R	·	·	·	·	·	·	·	·	·	·	·	·	·	·	·	·	·	·	NR	NR	R	NR	NR	NR
Chloracetic Acid, 50%	R	R	·	·	·	·	·	·	·	·	·	·	·	R	·	·	R	·	·	NR	NR	R	R	NR	NR
Chloral Hydrate	R	R	R	R	R	R	NR	NR	R	·	·	NR	NR	·	·	·	·	·	·	·	·	·	·	NR	NR
Chloramine	R	·	·	·	·	·	·	·	·	·	·	·	·	·	·	·	·	·	·	·	·	·	·	·	·
Chloric Acid, 20%	R	R	R	R	R	NR	NR	NR	R	R	·	NR	NR	·	·	·	·	·	·	·	·	·	·	·	·
Chlorine Gas, Dry	NR	NR	NR	NR	NR	NR	NR	NR	R	R	·	R	·	R	R	·	R	R	·	NR	NR	NR	NR	R	R
Chlorine Gas, Wet	NR	NR	NR	NR	NR	NR	NR	NR	R	R	·	R	·	NR	NR	NR	R	·	·	NR	NR	NR	NR	R	R
Chlorine Liquid	NR	NR	NR	NR	NR	NR	NR	NR	R	R	·	·	NR	·	·	·	·	·	·	NR	NR	NR	NR	R	R
Chlorine Water, Saturated	R	R	R	R	R	R	R	R	R	R	R	R	R	NR	NR	·	R	R	·	NR	NR	NR	NR	R	R
Chloracetic Acid, Conc.	NR	NR	R	·	NR	R	NR	NR	R	R	R	·	·	·	·	·	·	·	·	NR	NR	R	R	NR	NR
Chloroacetyl Chloride	R	·	·	·	·	·	·	·	·	·	·	·	·	·	·	·	·	·	·	·	·	·	·	·	·
Chlorobenzene	NR	NR	NR	NR	NR	NR	NR	NR	R	R	NR	NR	NR	R	R	·	R	·	·	NR	NR	NR	NR	R	R
Chlorobenzyl Chloride	NR	NR	NR	NR	NR	NR	NR	NR	R	·	·	·	·	·	·	·	·	·	·	·	·	·	·	R	R
Chloroform	NR	NR	NR	NR	NR	NR	NR	NR	R	R	·	NR	NR	R	R	·	NR	NR	NR	NR	NR	NR	NR	R	R
Chloropicrin	NR	NR	·	·	·	·	·	·	·	·	·	·	·	·	·	·	·	·	·	·	·	·	·	·	·
Chlorosulfonic Acid	·	NR	·	NR	NR	NR	NR	NR	·	NR	NR	NR	NR	NR	NR	NR	NR	NR	NR	NR	NR	NR	NR	NR	NR
Chlorox Bleach, 5.5% Cl₂	R	R	R	R	R	·	R	NR	R	R	·	NR	NR	·	·	·	·	·	·	R	R	R	NR	R	R
Chromic Acid, 10%	NR	NR	NR	NR	NR	R	R	NR	R	R	·	R	R	R	R	NR	R	R	·	NR	NR	NR	NR	R	R

Notes—Designations used are: R = Recommended; NR = Not Recommended; · = No data. All temperatures are in °F, and all chemicals are in solution in the pure or concentrated state unless indicated otherwise.

Chemical	PVC 70	PVC 140	CPVC 70	CPVC 140	CPVC 185	Polypropylene 70	Polypropylene 150	Polypropylene 180	PVDF 70	PVDF 150	PVDF 250	Polyethylene 70	Polyethylene 140	Fiberglass-Reinforced Epoxy 70	Fiberglass-Reinforced Epoxy 150	Fiberglass-Reinforced Epoxy 250	Fiberglass-Reinforced Vinyl/Polyester 70	Fiberglass-Reinforced Vinyl/Polyester 150	Fiberglass-Reinforced Vinyl/Polyester 250	Neoprene 70	Neoprene 185	EPDM 70	EPDM 185	Viton 70	Viton 185
Chromic Acid, 30%	NR	NR	NR	NR	NR	R	R	NR	R	R	·	R	R	NR	NR	NR	R	R	·	NR	NR	NR	NR	R	R
Chromic Acid, 40%	NR	NR	NR	NR	NR	R	R	NR	R	R	·	R	R	NR	NR	NR	R	R	·	NR	NR	NR	NR	R	R
Chromic Acid, 50%	NR	NR	NR	NR	NR	R	R	NR	R	R	NR	R	R	NR	NR	NR	R	R	·	NR	NR	NR	NR	R	R
Citric Acid	R	R	R	R	R	R	R	·	R	R	R	R	R	R	R	R	R	R	R	R	R	R	R	R	R
Coconut Oil	R	R	R	R	·	R	R	·	R	R	R	R	R	·	·	·	·	·	·	NR	NR	R	NR	R	R
Coke Oven Gas	NR	NR	R	R	R	R	·	·	R	R	·	·	·	·	·	·	·	·	·	NR	NR	NR	NR	R	R
Copper Carbonate	R	R	R	R	R	R	R	R	R	R	R	·	·	·	·	·	·	·	·	·	·	R	R	R	R
Copper Chloride	R	R	R	R	R	R	R	·	R	R	R	R	R	R	R	R	R	R	·	R	NR	R	R	R	R
Copper Cyanide	R	R	R	R	R	R	R	·	R	R	R	R	R	R	R	·	R	R	·	R	R	R	R	R	R
Copper Fluoride	R	R	R	R	R	R	R	·	R	R	R	R	R	R	R	R	R	R	·	·	·	R	R	R	R
Copper Nitrate	R	R	R	R	R	R	R	·	R	R	R	R	R	R	R	·	R	R	·	·	·	R	R	R	R
Copper Salts	R	R	R	R	R	R	R	·	R	R	R	R	R	R	R	·	R	R	·	·	·	R	R	R	R
Copper Sulfate	R	R	R	R	R	R	R	·	R	R	R	R	R	R	R	·	R	R	·	R	R	R	R	R	R
Corn Syrup	R	R	·	·	·	R	R	·	·	·	·	·	·	·	·	·	·	·	·	NR	NR	NR	NR	R	R
Cottonseed Oil	R	R	R	R	R	R	R	·	R	R	R	R	R	·	·	·	·	·	·	R	NR	R	NR	R	R
Cresol	NR	NR	·	NR	NR	R	NR	NR	R	R	·	NR	NR	·	·	·	·	·	·	NR	NR	NR	NR	R	R
Cresylic Acid, 50%	NR	NR	·	NR	NR	R	NR	NR	R	R	·	NR	NR	NR	NR	NR	NR	NR	NR	NR	NR	NR	NR	R	R
Croton Aldehyde	NR	NR	NR	NR	NR	R	·	·	R	R	NR	R	R	·	·	·	·	·	·	·	·	·	·	·	·
Crude Oil, Sour & Sweet	R	R	R	R	R	R	·	·	R	R	R	NR	NR	R	R	·	·	·	·	·	·	NR	NR	R	R
Cupric Fluoride	R	R	·	·	·	R	·	·	·	·	·	·	·	·	·	·	·	·	·	·	·	R	R	·	·
Cupric Sulfate	R	R	·	·	·	R	·	·	·	·	·	·	·	·	·	·	·	·	·	·	·	R	R	R	R
Cuprous Chloride	R	R	·	·	·	·	·	·	·	·	·	·	·	·	·	·	·	·	·	·	·	·	·	R	R
Cyclohexane	R	R	R	·	·	NR	NR	NR	R	R	R	R	R	·	·	·	R	·	·	NR	NR	NR	NR	R	R
Cyclohexanol	NR	NR	NR	NR	NR	R	NR	NR	R	R	NR	R	R	·	·	·	·	·	·	R	R	NR	NR	R	R
Cyclohexanone	NR	NR	NR	NR	NR	R	NR	NR	R	NR	NR	·	NR	·	·	·	·	·	·	NR	NR	R	NR	NR	NR
Decalin	NR	NR	NR	NR	NR	R	R	R	R	R	R	·	·	·	·	·	·	·	·	NR	NR	NR	NR	R	R
Desocyephedrine Hydrochloride	R	·	·	·	·	·	·	·	·	·	·	·	·	·	·	·	·	·	·	·	·	·	·	·	·
Detergents	R	R	R	R	R	R	R	·	R	R	R	R	R	R	R	·	R	R	·	R	NR	R	R	R	R
Detergent Solution (Heavy Duty)	R	R	R	R	R	R	R	·	R	R	R	·	·	R	R	·	R	R	·	R	NR	R	R	R	R
Dextrin	R	R	R	R	R	R	R	·	R	R	·	R	R	·	·	·	·	·	·	·	·	NR	NR	·	·
Dextrose	R	R	R	R	R	R	·	·	R	R	R	R	R	·	·	·	·	·	·	·	·	·	·	R	R
Diacetone Alcohol	NR	NR	NR	NR	NR	R	·	·	R	NR	NR	·	·	·	·	·	·	·	·	R	NR	R	R	NR	NR
Diazo Salts	R	R	R	R	R	R	R	·	R	·	·	R	R	·	·	·	·	·	·	·	·	·	·	·	·
Dibutoxy Ethyl Phthalate	NR	NR				·	·	·	·	·	·	·	·	·	·	·	·	·	·	·	·	·	·	·	·
Dibutyl Phthalate	NR	NR	NR	NR	NR	R	NR	NR	R	·	·	R	·	R	·	·	R	R	·	NR	NR	NR	NR	NR	NR
Dibutyl Sebacate	R	·	·	·	·	·	·	·	·	·	·	·	·	·	·	·	·	·	·	NR	NR	NR	NR	NR	NR
Dichlorobenzene (Ortho)	NR	NR	·	·	·	·	·	·	·	·	·	·	·	R	·	·	·	·	·	NR	NR	NR	NR	R	R
Dichloroethylene	NR	NR	NR	NR	NR	NR	NR	NR	R	·	·	·	·	R	·	·	·	·	·	NR	NR	NR	NR	R	R
Diesel Fuels	R	·	R	·	·	R	·	·	R	R	R	R	R	R	R	·	R	R	·	NR	NR	NR	NR	R	R
Diethylamine	NR	NR	·	·	·	·	·	·	R	NR	NR	·	·	·	·	·	·	·	·	R	·	NR	NR	NR	NR
Diethyl Cellosolve	·	·	·	·	·	·	·	·	R	R	R	·	·	·	·	·	·	·	·	·	·	NR	NR	·	·
Diethyl Ether	NR	NR	NR	NR	NR	R	NR	NR	R	NR	NR	·	·	NR	NR	NR	·	·	·	NR	NR	NR	NR	NR	NR
Diglycolic Acid	R	R	R	R	R	R	·	·	R	·	·	R	R	·	·	·	·	·	·	·	·	·	·	·	·
Dimethylamine	NR	NR	NR	NR	NR	R	·	·	NR	NR	NR	NR	NR	NR	NR	NR	·	·	·	R	NR	R	NR	NR	NR
Dimemthyl Formamide	NR	NR	NR	NR	NR	R	NR	NR	R	·	·	R	R	·	·	·	·	·	·	NR	NR	NR	NR	NR	NR
Dimethyl Hydrazine	NR	NR	·	·	·	·	·	·	·	·	·	·	·	·	·	·	·	·	·	·	·	·	·	NR	NR
Dicotyl Phthalate	NR	NR	NR	NR	NR	NR	NR	NR	R	·	·	NR	NR	·	·	·	R	·	·	·	·	·	·	·	·
Dioxane	NR	NR	NR	NR	NR	R	NR	NR	NR	NR	NR	·	·	R	·	·	·	·	·	NR	NR	R	NR	NR	NR
Dioxane, 1,4	NR	NR	NR	NR	NR	R	·	·	NR	NR	NR	·	·	·	·	·	·	·	·	·	·	·	·	NR	NR
Disodium Phosphate	R	R	R	R	R	R	R	R	R	R	R	R	R	·	·	·	·	·	·	·	·	R	R	·	·
Divinylbenzene	NR	NR	NR	NR	NR	NR	NR	NR	NR	NR	NR	·	·	R	R	·	·	·	·	NR	NR	NR	NR	R	R
Dowfax 9N9	·	·	·	·	·	·	·	·	·	·	·	·	·	R	R	·	·	·	·	NR	NR	NR	NR	R	R
Epsom Salt	R	·	·	·	·	·	·	·	·	·	·	·	·	·	·	·	·	·	·	·	·	R	R	·	·
Esters	NR	NR	NR	NR	NR	R	·	·	R	·	·	·	·	·	·	·	·	·	·	·	·	·	·	·	·
Ethanol	R	R	R	R	R	R	R	R	R	R	R	·	·	·	·	·	·	·	·	·	·	·	·	·	·
Ethers	NR	NR	NR	NR	NR	·	NR	NR	R	·	·	·	·	·	·	·	·	·	·	·	·	·	·	·	·
Ethyl Acetate	NR	NR	NR	NR	NR	R	NR	NR	R	NR	NR	NR	NR	R	R	·	R	·	·	NR	NR	R	NR	NR	NR
Ethyl Acetoacetate	NR	NR	NR	NR	NR	NR	NR	NR	R	NR	NR	·	·	·	·	·	R	·	·	NR	NR	R	NR	NR	NR
Ethyl Acrylate	NR	NR	NR	NR	NR	·	·	·	R	NR	NR	·	·	R	·	·	·	·	·	NR	NR	R	NR	NR	NR
Ethyl Alcohol	R	R	R	R	·	R	R	R	R	R	R	R	R	R	R	·	R	·	·	R	R	R	R	NR	NR
Ethyl Chloride	NR	NR	NR	NR	NR	R	NR	NR	R	R	R	NR	NR	·	·	·	R	·	·	NR	NR	NR	NR	R	R
Ethyl Chloroacetate	NR	NR	NR	NR	NR				·	·	·	·	·	·	·	·	·	·	·	·	·	·	·	·	·
Ethyl Ether	NR	NR	NR	NR	NR	NR	NR	NR	R	NR	NR	NR	NR	R	·	·	NR	NR	NR	NR	NR	NR	NR	NR	NR

Notes—Designations used are: R = Recommended; NR = Not Recommended; · = No data. All temperatures are in °F, and all chemicals are in solution in the pure or concentrated state unless indicated otherwise.

Chemical	PVC 70	PVC 140	CPVC 70	CPVC 140	CPVC 185	Polypropylene 70	Polypropylene 150	Polypropylene 180	PVDF 70	PVDF 150	PVDF 250	Polyethylene 70	Polyethylene 140	Fiberglass-Reinforced Epoxy 70	Fiberglass-Reinforced Epoxy 150	Fiberglass-Reinforced Epoxy 250	Fiberglass-Reinforced Vinyl/Polyester 70	Fiberglass-Reinforced Vinyl/Polyester 150	Fiberglass-Reinforced Vinyl/Polyester 250	Neoprene 70	Neoprene 185	EPDM 70	EPDM 185	Viton® 70	Viton® 185
Ethylene Bromide	NR	NR	NR	NR	NR	NR	NR	NR	R	R	R	NR	NR	·	·	·	·	·	·	·	·	NR	NR	·	·
Ethylene Chloride	NR	NR	NR	NR	NR	R	NR	NR	R	R	R	NR	NR	R	·	·	R	·	·	NR	NR	NR	NR	R	NR
Ethylene Chlorhydrin	NR	NR	NR	NR	NR	R	NR	NR	R	NR	NR	NR	NR	·	·	·	R	R	·	R	NR	R	NR	R	R
Ethylene Diamine	NR	NR	NR	NR	NR	R	·	·	·	NR	NR	·	·	NR	NR	NR	NR	NR	NR	R	R	R	R	NR	NR
Ethylene Dichloride	NR	NR	NR	NR	NR	R	NR	NR	R	R	R	NR	NR	NR	NR	NR	NR	NR	NR	NR	NR	NR	NR	R	R
Ethylene Glycol	R	R	R	R	R	R	R	R	R	R	R	R	R	R	R	R	R	R	·	R	R	R	R	R	R
Ethylene Oxide	NR	NR	NR	NR	NR	NR	NR	NR	R	R	·	R	·	NR	NR	NR	·	·	·	NR	NR	NR	NR	NR	NR
Fatty Acids	R	R	R	R	R	R	R	·	R	R	R	NR	NR	R	R	R	R	R	R	R	NR	NR	NR	R	R
Ferric Acetate	R	NR	·	·	·	·	·	·	·	·	·	·	·	·	·	·	·	·	·	·	·	·	·	·	·
Ferric Chloride	R	R	R	R	R	R	R	R	R	R	R	R	R	R	R	R	R	R	R	R	R	R	R	R	R
Ferric Hydroxide	R	R	R	R	R	·	·	·	·	·	·	R	R	·	·	·	·	·	·	·	·	·	·	·	·
Ferric Nitrate	R	R	R	R	R	R	R	R	R	R	R	R	R	R	R	R	R	R	R	R	·	R	R	R	R
Ferric Sulfate	R	R	R	R	R	R	R	R	R	R	R	R	·	R	R	·	R	R	R	R	R	R	R	R	R
Ferrous Chloride	R	R	R	R	R	R	R	R	R	R	R	R	R	R	R	·	R	R	·	·	·	·	·	·	·
Ferrous Hydroxide	R	·	·	·	·	·	·	·	·	·	·	·	·	·	·	·	·	·	·	·	·	·	·	·	·
Ferrous Nitrate	R	R	R	R	R	R	R	R	R	R	R	R	R	·	·	·	R	R	·	·	·	R	R	R	R
Ferrous Sulfate	R	R	R	R	R	R	R	R	R	R	R	R	R	R	R	·	R	R	R	·	·	R	R	R	R
Fish Solubles	R	R	R	R	R	R	·	·	R	R	R	R	R	·	·	·	·	·	·	NR	NR	NR	NR	R	R
Fluorine Gas, Wet	R	R	R	·	·	NR	NR	NR	R	·	·	NR	NR	R	·	·	R	·	·	NR	NR	NR	NR	NR	NR
Fluoboric Acid	R	R	R	R	NR	R	·	·	R	·	·	R	R	NR	NR	NR	R	R	·	R	R	R	R	·	·
Fluoric Acid	R	R	R	R	NR	·	·	·	·	·	·	R	R	R	R	·	R	R	R	·	·	·	·	·	·
Fluosilicic Acid	NR	NR	R	R	NR	R	·	·	R	R	R	R	R	R	R	·	R	R	·	R	NR	R	NR	NR	NR
Formaldehyde, 35%	R	R	R	R	NR	R	R	·	R	R	R	R	·	R	R	·	R	R	R	R	NR	R	R	NR	NR
Formaldehyde, 37%	R	R	R	R	NR	R	R	R	R	·	·	R	·	R	R	·	R	R	R	R	NR	R	NR	NR	NR
Formaldehyde, 50%	R	R	R	R	NR	R	R	·	R	·	·	·	·	·	·	·	R	R	·	R	NR	R	NR	NR	NR
Formic Acid (Anhydrous)	·	NR	R	NR	NR	R	R	NR	R	R	R	R	R	·	·	·	·	·	·	R	R	R	R	NR	NR
Formic Acid	R	NR	R	R	NR	R	R	·	R	R	R	R	R	R	R	·	R	·	·	R	R	R	R	NR	NR
Freon F-11	R	R	R	R	·	R	·	·	R	R	·	·	·	R	·	·	·	·	·	NR	NR	NR	NR	R	R
Freon F-12	NR	NR	R	·	·	R	·	·	R	R	·	·	·	R	·	·	·	·	·	R	R	R	R	R	NR
Freon F-21	NR	NR	·	·	·	·	·	·	R	·	·	·	·	·	·	·	·	·	·	NR	NR	NR	NR	NR	NR
Freon F-22	NR	NR	NR	NR	NR	R	·	·	R	·	·	·	·	R	·	·	·	·	·	R	R	R	R	NR	NR
Freon F-113	R	·	R	·	·	R	·	·	R	·	·	·	·	·	·	·	·	·	·	R	R	NR	NR	NR	NR
Freon F-114	R	·	R	·	·	R	·	·	R	·	·	·	·	·	·	·	·	·	·	R	R	R	R	R	NR
Fructose	R	R	R	R	R	R	R	·	R	R	R	R	R	·	·	·	·	·	·	·	·	·	·	R	R
Fruit Juices, Pulp	R	R	R	R	R	R	R	·	R	R	R	R	R	·	·	·	·	·	·	·	·	·	·	R	R
Fuel Oil	R	R	·	·	·	R	NR	NR	R	R	R	NR	NR	R	R	·	R	R	·	R	NR	NR	NR	R	R
Furfural	NR	NR	NR	NR	NR	NR	NR	NR	R	NR	NR	NR	NR	NR	NR	NR	NR	NR	NR	NR	NR	R	NR	NR	NR
Gallic Acid	R	R	R	R	·	R	·	·	R	·	·	R	R	·	·	·	·	·	·	R	NR	R	NR	R	R
Gas, Manufactured	R	R	R	R	R	R	R	·	R	R	R	·	·	·	·	·	·	·	·	·	·	·	·	R	R
Gas, Natural	R	R	R	R	R	R	·	·	R	R	R	·	·	R	R	·	R	R	·	·	·	·	·	R	R
Gasoline, Leaded	·	·	R	·	NR	NR	NR	NR	R	R	R	NR	NR	R	R	·	R	R	R	NR	NR	NR	NR	R	R
Gasoline, Unleaded	·	·	R	·	NR	NR	NR	NR	R	R	R	NR	NR	R	R	·	R	R	R	NR	NR	NR	NR	R	R
Gasoline, Refined	·	·	R	·	NR	NR	NR	NR	R	R	R	NR	NR	R	R	·	R	R	·	NR	NR	NR	NR	R	R
Gasoline, Sour	R	R	·	·	·	NR	NR	NR	R	R	R	NR	NR	R	R	R	R	R	·	NR	NR	NR	NR	R	R
Gelatin	R	R	R	R	R	R	R	R	R	R	R	R	R	·	·	·	·	·	·	R	R	R	R	R	R
Gin	R	R	R	R	R	R	·	·	R	R	R	·	NR	·	·	·	·	·	·	·	·	·	·	·	·
Glucose	R	R	R	R	R	R	R	R	R	R	R	R	R	R	R	R	·	·	·	R	R	R	R	R	R
Glue	R	R	R	R	R	R	·	·	R	R	·	·	·	·	·	·	·	·	·	R	R	R	R	R	R
Glycerine, Clycerol	R	R	R	R	R	R	R	R	R	R	R	R	R	R	R	·	R	R	R	·	·	R	R	R	R
Glycolic	R	R	R	R	R	R	R	·	R	NR	NR	R	·	NR	NR	NR	R	·	·	·	·	·	·	·	·
Glycols	R	R	R	R	R	R	NR	NR	R	R	R	R	R	R	R	·	·	·	·	R	R	R	R	R	R
Green Liquor	R	R	R	R	R	R	R	·	R	R	·	R	R	R	R	·	·	·	NR	R	NR	R	R	R	R
Grape Sugar	R	R	·	·	·	·	·	·	·	·	·	R	R	·	·	·	·	·	·	·	·	·	·	R	R
Heptane	R	NR	R	R	·	R	NR	NR	R	R	R	NR	NR	R	R	·	R	R	·	·	·	NR	NR	R	R
Hexane	NR	NR	R	·	·	R	NR	NR	R	R	R	NR	NR	R	R	·	R	·	·	NR	NR	R	NR	R	R
Hydrobromic Acid, 20%	R	R	R	R	NR	R	R	·	R	R	R	R	R	R	R	·	R	·	·	NR	NR	R	R	R	R
Hydrobromic Acid, 50%	·	NR	·	·	·	R	NR	NR	R	R	R	·	·	R	R	·	R	·	·	R	R	R	R	R	R
Hydrochloric Acid, 20%	R	R	R	R	R	R	·	NR	R	R	R	R	R	R	R	·	R	·	·	R	NR	R	R	R	R
Hydrochloric Acid, 38%	R	R	R	R	R	R	NR	NR	R	R	R	R	R	R	R	·	R	·	·	NR	NR	NR	NR	R	NR
Hydrocyanic Acid	R	R	R	R	R	R	R	·	R	R	R	R	R	R	·	·	R	·	·	R	NR	R	R	R	R
Hydrocyanic Acid, 10%	R	R	R	R	R	R	·	·	R	R	R	R	R	R	·	·	R	R	·	·	·	R	R	R	R

Notes—Designations used are: R = Recommended; NR = Not Recommended; · = No data. All temperatures are in °F, and all chemicals are in solution in the pure or concentrated state unless indicated otherwise.

	PVC		CPVC			Polypro-pylene			PVDF			Poly-eth-ylene		Fiber-glass-Rein-forced Epoxy			Fiber-glass-Rein-forced Vinyl/ Polyester			Neo-prene		EPDM		Viton®	
Chemical	70	140	70	140	185	70	150	180	70	150	250	70	140	70	150	250	70	150	250	70	185	70	185	70	185
Hydrofluoric Acid, Dilute	R	NR	NR	NR	NR	R	R	·	R	R	R	R	R	·	·	·	R	R	·	NR	NR	R	NR	R	NR
Hydrofluoric Acid, 30%	R	NR	NR	NR	NR	R	·	·	R	R	R	R	R	NR	NR	NR	NR	NR	NR	NR	NR	R	NR	R	NR
Hydrofluoric Acid, 40%	R	NR	NR	NR	NR	R	·	NR	R	R	R	R	R	NR	NR	NR	NR	NR	NR	NR	NR	R	NR	R	NR
Hydrofluoric Acid, 50%	R	NR	NR	NR	NR	R	·	·	R	R	R	R	R	NR	NR	NR	NR	NR	NR	NR	NR	R	NR	R	NR
Hydrofluosilicic Acid	NR	NR	NR	NR	NR	R	·	·	R	R	R	R	R	R	·	·	R	·	·	R	NR	R	NR	R	R
Hydrogen	R	R	R	R	·	R	·	·	R	R	R	R	R	R	R	·	R	R	·	R	R	R	R	R	R
Hydrogen Cyanide	R	R	R	R	R	R	·	·	R	R	R	·	·	·	·	·	·	·	·	·	·	·	·	·	·
Hydrogen Fluoride, Anhydros	NR	NR	NR	NR	NR	R	·	·	R	R	·	·	·	·	·	·	·	·	·	·	·	·	·	NR	NR
Hydrogen Peroxide, Dilute	R	R	R	R	·	R	·	·	R	R	·	R	R	R	·	·	R	·	·	NR	NR	NR	NR	R	R
Hydrogen Peroxide, 30%	R	R	R	R	·	R	NR	NR	R	R	R	R	R	R	·	NR	R	·	·	NR	NR	NR	NR	R	R
Hydrogen Peroxide, 50%	R	R	R	R	·	R	NR	NR	R	·	·	R	R	NR	NR	NR	·	·	·	NR	NR	NR	NR	R	R
Hydrogen Phosphide	NR	NR	R	·	·	R	·	·	R	·	·	R	R	·	·	·	·	·	·	·	·	·	·	·	·
Hydrogen Sulfide Dry	R	R	R	R	R	R	R	·	R	R	R	R	R	R	R	·	R	R	·	R	NR	R	R	NR	NR
Hydrogen Sulfide Ag Sol	R	R	R	R	R	R	·	·	R	R	R	R	R	R	R	·	R	R	·	NR	NR	R	R	NR	NR
Hydroquinone	R	R	R	R	R	R	R	·	R	R	·	R	R	·	·	·	·	·	·	·	·	NR	NR	R	R
Hydroxylamine Sulfate	R	R	R	R	R	R	R	·	R	R	·	·	·	·	·	·	·	·	·	·	·	·	·	·	·
Hypochlorous Acid, 10%	R	R	R	R	·	R	·	·	R	R	R	R	R	R	·	·	R	·	·	NR	NR	R	NR	R	R
Hypochlorous Acid, 50%	R	R	R	R	·	R	R	·	R	R	R	R	R	R	·	·	R	·	·	NR	NR	R	NR	R	R
Hydrazine	NR	NR	·	·	·	·	·	·	·	·	·	·	·	·	·	·	·	·	·	·	·	·	·	·	·
Iodine	NR	NR	R	NR	·	NR	NR	NR	R	R	·	R	·	NR	NR	NR	·	·	·	·	·	R	NR	R	R
Iodine Solution 10%	NR	NR	R	·	·	R	·	·	R	R	·	·	·	R	R	·	·	·	·	·	·	R	NR	R	R
Isooctane	R	NR	R	NR	NR	R	NR	NR	R	R	R	·	·	·	·	·	·	·	·	R	NR	NR	NR	R	R
Isopropyl Alcohol	R	·	R	·	·	R	R	R	R	R	·	R	R	R	·	·	·	·	·	R	NR	R	R	R	R
Isopropyl Ether	NR	NR	NR	NR	NR	·	NR	NR	R	R	·	·	·	·	·	·	·	·	·	NR	NR	NR	NR	NR	NR
Jet Fuel, JP 4	R	R	R	R	R	·	NR	NR	R	R	·	·	·	R	R	·	R	R	·	NR	NR	NR	NR	R	R
Jet Fuel, JP 5	R	R	R	R	R	·	NR	NR	R	R	·	·	·	R	R	·	R	R	·	NR	NR	NR	NR	R	R
Kerosene	R	R	R	R	R	R	NR	NR	R	R	R	NR	NR	R	R	R	R	R	·	R	NR	NR	NR	R	R
Ketones	NR	NR	NR	NR	NR	R	NR	NR	R	NR	NR	NR	NR	·	·	·	·	·	·	·	·	·	·	·	·
Kraft Liquor	R	R	R	R	R	R	R	·	R	·	·	R	R	·	·	·	·	·	·	·	·	·	·	·	·
Lactic Acid, 25%	R	R	R	R	R	R	R	·	R	NR	NR	R	R	R	R	·	R	R	·	R	NR	R	NR	R	R
Lard Oil	R	R	R	R	R	R	·	·	R	R	R	·	·	·	·	·	·	·	·	NR	NR	NR	NR	R	R
Lauric Acid	R	R	R	R	R	R	R	·	R	R	·	·	·	R	R	·	R	R	·	·	·	·	·	·	·
Lauryl Chloride	R	R	R	R	·	R	·	·	R	R	R	R	NR	·	·	·	R	R	·	·	·	·	·	·	·
Lead Acetate	R	R	R	R	R	R	R	R	R	R	R	R	R	R	R	R	R	R	R	R	NR	R	R	NR	NR
Lead Chloride	R	R	R	R	R	·	·	·	·	·	·	·	·	R	R	·	R	R	·	·	·	R	R	·	·
Lead Nitrate	R	R	·	·	·	·	·	·	·	·	·	·	·	·	·	·	·	·	·	R	R	R	R	R	R
Lead Sulfate	R	R	R	R	R	·	·	·	·	·	·	·	·	·	·	·	·	·	·	·	·	R	R	·	·
Lemon Oil	R	·	R	·	·	·	NR	NR	R	R	R	·	·	·	·	·	·	·	·	·	·	·	·	·	·
Ligroine	NR	NR	NR	NR	NR	R	·	·	R	NR	NR	·	·	·	·	·	·	·	·	R	NR	NR	NR	R	R
Lime Sulfur	R	R	R	R	·	R	R	·	R	R	·	R	R	·	·	·	·	·	·	R	R	R	R	R	R
Linoleic Acid	R	R	R	R	R	R	R	R	R	R	R	·	·	·	·	·	·	·	·	R	R	R	R	R	NR
Linoleic Oil	R	R	R	R	R	R	·	·	R	R	·	·	·	·	·	·	·	·	·	·	·	·	·	·	·
Linseed Oil	R	R	R	R	R	R	R	·	R	R	R	·	NR	R	R	·	R	R	·	R	NR	NR	NR	R	R
Liqueurs	R	R	R	R	·	R	R	·	R	R	R	R	R	·	·	·	·	·	·	·	·	·	·	R	R
Lithium Bromide	R	R	R	R	R	·	·	·	·	·	·	·	·	·	·	·	·	·	·	·	·	·	·	·	·
Lithium Chloride, Saturated	·	·	·	·	·	·	·	·	·	·	·	·	·	R	R	·	·	·	·	·	·	·	·	·	·
Lithium Hydroxide, Saturated	·	·	·	·	·	·	·	·	·	·	·	·	·	R	R	·	·	·	·	·	·	·	·	·	·
Lubricating Oil	R	R	R	R	R	R	NR	NR	R	R	R	·	NR	·	·	·	·	·	·	R	NR	NR	NR	R	R
Lye	R	R	R	R	R	R	·	·	·	·	·	·	·	·	·	·	·	·	·	R	NR	R	R	R	NR
Machine Oil	R	R	R	R	R	R	NR	NR	R	R	·	·	·	·	·	·	·	·	·	·	·	NR	NR	·	·
Magnesium Carbonate	R	R	R	R	R	R	R	·	R	R	R	R	R	R	R	R	R	R	R	·	·	R	R	R	·
Magnesium Chloride	R	R	R	R	R	R	R	R	R	R	R	R	R	R	R	R	R	R	R	R	R	R	R	R	R
Maleic Acid	R	R	R	R	R	R	R	·	R	R	R	R	R	R	R	·	R	R	R	NR	NR	R	NR	R	R
Malic Acid	R	R	R	R	R	R	NR	NR	R	R	R	R	R	·	·	·	·	·	·	NR	NR	R	NR	R	R
Manganese Salts	R	·	R	R	·	R	R	R	R	R	·	R	R	·	·	·	·	·	·	·	·	·	·	·	·
Manganese Sulfate	R	·	·	·	·	R	R	·	R	R	R	·	·	·	·	·	·	·	·	·	·	·	·	·	·
Manganese Sulfate	R	·	·	·	·	R	R	·	R	R	R	·	·	·	·	·	·	·	·	·	·	·	·	·	·
Maleic Acid	R	R	R	R	R	R	R	·	R	R	R	R	R	R	R	·	R	R	R	NR	NR	R	NR	R	R
Malic Acid	R	R	R	R	R	R	NR	NR	R	R	R	R	R	·	·	·	·	·	·	NR	NR	R	NR	R	R
Manganese Salts	R	·	R	R	·	R	R	R	R	R	·	R	R	·	·	·	·	·	·	·	·	·	·	·	·

Notes—Designations used are: R = Recommended; NR = Not Recommended; · = No data. All temperatures are in °F, and all chemicals are in solution in the pure or concentrated state unless indicated otherwise.

IN DEGREES FAHRENHEIT

Chemical	PVC 70	PVC 140	CPVC 70	CPVC 140	CPVC 185	Polypropylene 70	Polypropylene 150	Polypropylene 180	PVDF 70	PVDF 150	PVDF 250	Polyethylene 70	Polyethylene 140	FRP Epoxy 70	FRP Epoxy 150	FRP Epoxy 250	FRP Vinyl/Polyester 70	FRP Vinyl/Polyester 150	FRP Vinyl/Polyester 250	Neoprene 70	Neoprene 185	EPDM 70	EPDM 185	Viton 70	Viton 185
Mercuric Chloride	R	R	R	R	R	R	R	R	R	R	R	R	R	R	R	R	R	R	·	R	R	R	R	R	R
Mercuric Cyanide	R	R	R	R	R	R	R	R	R	R	R	R	R	·	·	·	·	·	·	·	·	R	R	·	·
Mercuric Sulfate	R	R	·	·	·	·	·	·	·	·	·	·	·	·	·	·	·	·	·	·	·	R	R	R	R
Mercurous Nitrate	R	R	R	R	R	·	·	·	·	·	·	R	R	·	·	·	·	·	·	·	·	·	·	·	·
Mercury	R	R	R	R	R	R	R	·	R	R	R	R	R	R	R	R	R	R	R	R	R	R	R	R	R
Methane	R	R	R	·	·	R	·	·	R	R	·	·	·	·	·	·	·	·	·	R	NR	NR	NR	R	NR
Methelene Chlorobromide	NR	NR	·	·	·	·	·	·	·	·	·	·	·	·	·	·	·	·	·	·	·	·	·	·	·
Methoxyethyl Oleate	R	·	·	·	·	·	·	·	·	·	·	·	·	·	·	·	·	·	·	·	·	·	·	·	·
Methylamine	NR	NR	·	·	·	·	·	·	·	·	·	·	·	·	·	·	·	·	·	·	·	·	·	·	·
Methyl Bromide	NR	NR	NR	NR	NR	NR	NR	NR	R	R	R	NR	NR	·	·	·	·	·	·	NR	NR	NR	NR	R	R
Methyl Cellosolve	NR	NR	NR	NR	NR	R	·	·	R	R	R	·	·	·	·	·	·	·	·	NR	NR	R	NR	NR	NR
Methyl Chloride	NR	NR	NR	NR	NR	NR	NR	NR	R	R	R	NR	NR	NR	NR	NR	R	·	·	NR	NR	NR	NR	R	NR
Methyl Chloroform	NR	NR	NR	NR	NR	·	·	·	R	·	·	·	·	·	·	·	·	·	·	·	·	·	·	·	·
Methyl Ethyl Ketone	NR	NR	NR	NR	NR	R	NR	NR	·	·	NR	NR	NR	R	·	·	NR	NR	NR	NR	NR	R	R	NR	NR
Methyl Isobutyl Ketone	NR	NR	NR	NR	NR	R	NR	NR	R	NR	NR	R	R	R	R	·	NR	NR	NR	NR	NR	R	NR	NR	NR
Methyl Methacrylate	R	·	R	·	·	·	·	·	R	·	·	·	·	NR	NR	NR	NR	NR	NR	NR	NR	NR	NR	NR	NR
Methyl Salicylate	R	·	R	·	·	R	·	·	R	·	·	·	·	·	·	·	·	·	·	NR	NR	R	NR	·	·
Methyl Sulfate	R	NR	R	NR	NR	R	NR	NR	R	·	·	·	·	·	·	·	·	·	·	·	·	·	·	·	·
Methyl Sulfuric Acid	R	R	R	R	R	R	·	·	R	·	·	R	R	·	·	·	·	·	·	·	·	·	·	·	·
Methylene Chloride	NR	NR	NR	NR	NR	NR	NR	NR	R	NR	NR	NR	NR	R	·	·	NR	NR	NR	·	·	NR	NR	R	NR
Methylene Iodine	R	·	·	·	·	·	·	·	·	·	·	·	·	·	·	·	·	·	·	·	·	·	·	·	·
Methylisobutyl Carbinol	R	·	R	·	·	R	·	·	R	R	·	·	·	·	·	·	·	·	·	·	·	·	·	·	·
Milk	R	R	R	R	·	R	R	R	R	R	·	R	R	·	·	·	·	·	·	R	R	R	R	R	R
Mineral Oil	R	R	R	R	R	R	NR	NR	R	R	R	R	NR	R	R	R	R	R	R	R	NR	NR	NR	R	R
Molasses	R	R	R	R	R	R	R	·	·	·	·	R	R	·	·	·	·	·	·	·	·	R	R	R	R
Monoethanolamine	·	·	·	·	·	R	R	·	·	NR	NR	·	·	·	·	·	·	·	·	·	·	R	NR	R	R
Motor Oil	R	R	R	R	R	·	R	NR	R	·	·	·	·	·	·	·	·	·	·	NR	NR	NR	NR	R	R
Naphtha	R	R	R	R	·	R	·	·	R	R	R	NR	NR	R	R	R	R	R	·	NR	NR	NR	NR	R	R
Napthalene	NR	NR	NR	NR	NR	R	R	R	R	R	R	R	R	R	R	·	R	R	R	NR	NR	NR	NR	R	R
Natural Gas	R	R	R	·	·	R	·	·	R	R	R	·	·	·	·	·	·	·	·	R	R	NR	NR	R	R
Nickel Acetate	R	·	·	·	·	·	·	·	·	·	·	·	·	·	·	·	·	·	·	R	NR	R	R	NR	NR
Nickel Chloride	R	R	R	R	R	R	R	R	R	R	R	R	R	R	R	R	R	R	R	R	R	R	R	R	R
Nickel Nitrate	R	R	R	R	R	R	R	R	R	R	R	R	R	R	R	R	R	R	R	R	R	R	R	R	R
Nickel Salt	R	R	R	R	R	R	R	R	R	R	R	R	R	R	R	·	R	R	·	R	R	R	R	R	R
Nickel Sulfate	R	R	R	R	R	R	R	R	R	R	R	R	R	R	R	R	R	R	R	R	R	R	R	R	R
Nicotine	R	R	R	R	R	R	R	·	R	R	·	R	R	·	·	·	·	·	·	·	·	·	·	·	·
Nicotinic Acid	R	R	R	R	R	R	·	·	R	R	R	R	R	·	·	·	·	·	·	·	·	·	·	·	·
Nitric Acid, 10%	R	NR	R	R	R	R	R	·	R	·	·	R	R	R	·	·	NR	NR	·	R	NR	R	NR	R	R
Nitric Acid, 30%	R	NR	R	R	R	R	R	·	R	·	·	R	·	R	·	NR	R	·	·	R	NR	R	NR	R	R
Nitric Acid, 40%	R	NR	R	NR	NR	R	R	·	R	·	·	NR	NR	NR	NR	NR	R	·	·	NR	NR	NR	NR	R	R
Nitric Acid, 50%	R	NR	R	NR	NR	R	NR	NR	R	·	·	NR	NR	NR	NR	NR	R	·	·	NR	NR	NR	NR	R	R
Nitric Acid, 70%	NR	NR	NR	NR	NR	NR	NR	NR	NR	·	·	NR	NR	NR	NR	NR	R	·	·	NR	NR	NR	NR	R	R
Nitric Acid, 100%	NR	NR	NR	NR	NR	NR	NR	NR	NR	·	·	NR	NR	NR	NR	NR	·	·	·	NR	NR	NR	NR	R	NR
Nitrobenzene	NR	NR	NR	NR	NR	NR	NR	NR	R	NR	NR	NR	NR	NR	NR	NR	NR	NR	NR	NR	NR	R	R	R	NR
Nitroglycerine	NR	NR	NR	NR	NR	R	·	·	·	·	·	·	·	·	·	·	·	·	·	·	·	·	·	·	·
Nitrous Acid, 10%	R	R	R	R	R	NR	NR	NR	R	R	·	·	·	NR	NR	NR	R	·	·	·	·	·	·	·	·
Nitrous Oxide	R	R	R	R	·	R	·	·	R	R	·	·	·	·	·	·	·	·	·	·	·	·	·	·	·
Nitroglycol	NR	NR	·	·	·	·	·	·	·	·	·	·	·	·	·	·	·	·	·	·	·	·	·	·	·
Ocenol	R	R	R	R	R	NR	NR	NR	R	·	·	·	·	·	·	·	·	·	·	·	·	·	·	·	·
Oil & Fats	R	R	R	R	R	R	·	·	R	R	R	R	NR	·	·	·	·	·	·	·	·	·	·	·	·
Oils, Vegetable	R	R	R	R	R	R	·	·	R	R	R	R	NR	R	R	R	·	·	·	·	·	NR	NR	R	R
Oleic Acid	R	R	R	R	R	R	NR	NR	R	R	R	R	NR	R	R	·	R	R	R	NR	NR	NR	NR	R	NR
Oleum	NR	NR	NR	NR	NR	NR	NR	NR	NR	NR	NR	NR	NR	·	·	·	·	·	·	NR	NR	NR	NR	R	R
Oxalic Acid, 50%	R	R	R	R	R	R	R	·	R	·	·	R	R	R	R	·	R	R	·	R	NR	R	R	R	R
Oxygen Gas	R	R	R	R	R	NR	NR	NR	R	R	R	NR	NR	·	·	·	·	·	·	R	R	R	R	R	R
Ozone	R	R	R	R	R	NR	NR	NR	R	R	R	NR	NR	·	·	·	·	·	·	NR	NR	R	R	R	R
Palmitic Acid	·	·	·	·	·	R	R	R	R	R	R	·	·	·	·	·	·	·	·	R	R	R	NR	R	R
Palmitic Acid, 10%	R	R	R	R	·	R	R	R	R	R	R	R	R	·	·	·	·	·	·	R	R	·	·	R	R
Palmitic Acid, 70%	NR	NR	R	R	·	R	R	R	R	R	R	R	R	·	·	·	·	·	·	R	R	·	·	R	·
Paraffin	R	R	·	·	·	·	·	·	·	·	·	·	·	·	·	·	·	·	·	·	·	·	·	R	R
Peracetic Acid, 40%	NR	NR	NR	NR	NR	NR	NR	NR	R	·	·	·	·	·	·	·	·	·	·	·	·	·	·	·	·
Perchlorethylene	·	·	·	·	·	NR	NR	NR	R	R	R	·	·	R	R	·	R	·	·	NR	NR	NR	NR	R	R

Notes—Designations used are: R = Recommended; NR = Not Recommended; · = No data. All temperatures are in °F, and all chemicals are in solution in the pure or concentrated state unless indicated otherwise.

	IN DEGREES FAHRENHEIT																								
	PVC		CPVC			Polypro-pylene			PVDF			Poly-eth-ylene		Fiber-glass-Rein-forced Epoxy			Fiber-glass-Rein-forced Vinyl/Polyester			Neo-prene		EPDM		Viton®	
Chemical	70	140	70	140	185	70	150	180	70	150	250	70	140	70	150	250	70	150	250	70	185	70	185	70	185
Perchloric Acid, 10%	R	·	R	R	·	R	R	R	R	R	·	R	R	R	·	·	R	·	·	R	NR	R	NR	R	R
Perchloric Acid, 70%	NR	NR	NR	NR	·	R	R	R	R	·	·	R	R	R	·	·	R	·	·	R	NR	R	NR	R	R
Perphosphate	R	·	·	·	·	·	·	·	·	·	·	·	·	·	·	·	·	·	·	·	·	·	·	·	·
Petrolatum	R	R	R	R	R	R	·	·	R	R	R	·	·	·	·	·	·	·	·	·	·	·	·	·	·
Petroleum Oils, Refined	·	·	·	·	·	R	·	·	·	·	·	·	·	·	·	·	·	·	·	R	NR	NR	NR	R	R
Petroleum Oils, Sour	R	R	R	R	R	R	NR	NR	R	R	R	·	·	·	·	·	·	·	·	R	NR	NR	NR	R	R
Phenol	R	NR	R	R	·	R	R	R	R	R	·	NR	NR	NR	NR	NR	NR	NR	NR	NR	NR	NR	NR	R	R
Phenylhydrazine	NR	NR	NR	NR	NR	NR	NR	NR	R	R	·	·	·	·	·	·	·	·	·	NR	NR	R	NR	R	R
Phenylhydrazine Hydrochloride	R	NR	R	R	R	NR	NR	NR	R	R	·	·	·	·	·	·	·	·	·	·	·	·	·	·	·
Phosgene Gas	NR	NR	NR	NR	NR	NR	NR	NR	R	·	·	·	·	·	·	·	·	·	·	·	·	·	·	NR	NR
Phosgene Liquid	NR	NR	NR	NR	NR	NR	NR	NR	·	·	·	·	·	·	·	·	·	·	·	·	·	R	NR	NR	NR
Phosphoric Acid, 10%	R	R	R	R	R	R	R	R	R	R	R	R	R	R	R	·	R	R	R	R	NR	R	R	R	R
Phosphoric Acid, 50%	R	R	R	R	R	R	R	R	R	R	·	R	R	R	R	·	R	R	R	R	NR	R	R	R	R
Phosphoric Acid, 85%	R	R	R	R	R	R	R	R	R	R	·	R	R	R	R	·	R	R	R	R	NR	R	NR	R	R
Phosphorus, Yellow	R	·	·	·	·	·	·	·	·	·	·	R	NR	·	·	·	·	·	·	·	·	·	·	·	·
Phosphorus, Pentoxide	R	NR	·	·	·	R	NR	NR	R	R	·	R	NR	R	R	·	R	R	·	·	·	R	·	·	·
Phosphorus, Trichloride	NR	NR	NR	NR	NR	NR	NR	NR	R	R	·	NR	·	·	·	·	·	·	·	NR	NR	R	R	R	R
Photographic Solutions	R	R	R	R	R	R	R	·	R	·	·	R	·	·	·	·	·	·	·	·	·	·	·	R	R
Picric Acid	NR	NR	NR	NR	NR	R	R	·	R	·	·	R	·	·	·	·	·	·	·	R	R	R	NR	R	R
Plating Solutions, Brass	R	R	R	R	R	R	R	·	R	·	·	R	R	·	·	·	·	·	·	NR	NR	R	R	R	R
Plating Solutions, Cadmium	R	R	R	R	R	R	R	·	R	·	·	R	R	·	·	·	·	·	·	NR	NR	R	R	R	R
Plating Solutions, Chrome	R	R	R	R	R	R	R	·	R	·	·	NR	NR	·	·	·	R	R	·	NR	NR	R	R	R	R
Plating Solutions, Copper	R	R	R	R	R	R	R	·	R	R	R	R	R	·	·	·	·	·	·	NR	NR	R	R	R	R
Plating Solutions, Gold	R	R	R	R	R	R	R	·	R	·	·	R	R	·	·	·	R	R	·	NR	NR	R	R	R	R
Plating Solutions, Lead	R	R	R	R	R	R	R	·	R	·	·	R	R	·	·	·	R	R	·	NR	NR	R	R	R	R
Plating Solutions, Nickel	R	R	R	R	R	R	R	·	R	R	R	R	R	R	R	·	R	R	·	NR	NR	R	R	R	R
Plating Solutions, Rhodium	R	R	R	R	R	R	R	·	R	·	·	R	R	·	·	·	·	·	·	NR	NR	·	·	R	R
Plating Solutions, Silver	R	R	R	R	R	R	R	·	R	R	R	R	R	R	R	·	R	R	·	NR	NR	R	R	R	R
Plating Solutions, Tin	R	R	R	R	R	R	R	·	R	R	R	R	R	·	·	·	·	·	·	NR	NR	R	R	R	R
Plating Solutions, Zinc	R	R	R	R	R	R	R	·	R	R	R	R	R	·	·	·	·	·	·	NR	NR	R	R	R	R
Polyvinyl Acetate	·	·	·	·	·	R	·	·	R	R	R	·	·	NR	NR	NR	R	R	·	R	R	R	R	R	R
Potash	R	R	·	·	·	·	·	·	·	·	·	·	·	·	·	·	·	·	·	·	·	R	·	·	·
Potassium Alum	R	R	·	·	·	·	·	·	·	·	·	·	·	·	·	·	·	·	·	·	·	R	R	·	·
Potassium Aluminum Sulfate	R	R	R	R	R	R	R	R	·	·	·	·	·	·	·	·	R	R	·	·	·	R	R	·	·
Potassium Amyl Xanthate	R	NR	·	·	·	·	·	·	·	·	·	·	·	·	·	·	·	·	·	·	·	·	·	·	·
Potassium Bicarbonate	R	R	R	R	NR	R	R	·	R	R	R	R	R	R	R	R	R	R	R	·	·	·	·	R	R
Potassium Bichromate	R	R	R	R	R	R	R	·	R	R	R	R	R	·	·	·	·	·	·	·	·	·	·	R	R
Potassium Bisulfate	R	R	·	·	·	·	·	·	·	·	·	R	R	·	·	·	·	·	·	·	·	·	·	R	R
Potassium Borate	R	R	R	R	R	R	·	·	·	·	·	R	R	·	·	·	·	·	·	·	·	R	R	·	·
Potassium Bromate	R	R	R	R	R	R	·	·	·	·	·	R	R	·	·	·	·	·	·	·	·	R	R	R	R
Potassium Bromide	R	R	R	R	R	R	R	R	R	R	R	R	R	R	R	·	·	·	·	·	·	R	R	R	R
Potassium Carbonate	R	R	R	R	R	R	R	R	R	R	R	R	R	R	R	·	R	·	·	·	·	R	R	R	R
Potassium Chlorate, Aqueous	R	R	R	R	R	R	R	R	R	R	R	R	R	·	·	·	·	·	·	·	·	R	NR	R	R
Potassium Chloride	R	R	R	R	R	R	R	R	R	R	R	R	R	R	R	R	R	R	R	R	R	R	R	R	R
Potassium Chromate	R	R	R	R	R	R	R	·	R	R	R	R	R	·	·	·	·	·	·	·	·	·	·	·	·
Potassium Chlorate	R	R	·	·	·	R	R	·	·	·	·	R	R	·	·	·	·	·	·	·	·	R	NR	R	R
Potassium Cyanide	R	R	R	R	R	R	R	R	R	R	R	R	R	R	R	·	·	·	·	R	R	R	R	R	R
Potassium Dichromate	R	R	R	R	R	R	R	R	R	R	R	R	R	R	R	·	R	R	R	R	R	R	R	R	R
Potassium Ethyl Xanthate	R	NR	·	·	·	·	·	·	·	·	·	·	·	·	·	·	·	·	·	·	·	·	·	·	·
Potassium Ferricyanide	R	R	R	R	R	R	·	·	R	R	R	R	R	R	R	·	R	R	·	·	·	R	R	R	R
Potassium Ferrocyanide	R	R	R	R	R	R	R	R	R	R	R	R	R	R	R	·	R	R	·	·	·	R	R	R	R
Potassium Fluoride	R	R	R	R	R	R	R	·	R	R	R	R	R	·	·	·	·	·	·	·	·	·	·	R	R
Potassium Hydroxide	R	R	R	R	R	R	R	R	R	R	·	R	R	R	R	·	NR	NR	NR	R	NR	R	R	NR	NR
Potassium Hydroxide, 25%	R	R	R	R	R	R	R	R	R	R	·	R	R	R	R	·	NR	NR	NR	R	NR	R	R	NR	NR
Potassium Hypochlorite	R	R	R	R	·	R	NR	NR	R	R	R	·	·	·	·	·	·	·	·	·	·	·	·	·	·
Potassium Iodide	R	·	R	·	·	R	·	·	R	R	R	·	·	·	·	·	·	·	·	·	·	R	·	·	·
Potassium Nitrate	R	R	R	R	R	R	R	R	R	R	R	R	R	R	R	R	R	R	R	R	R	R	R	R	R
Potassium Perborate	R	R	R	R	R	R	R	·	R	R	·	R	R	·	·	·	·	·	·	·	·	R	R	·	·
Potassium Perchlorate	R	R	R	R	R	R	R	·	R	R	·	·	·	·	·	·	·	·	·	·	·	R	·	·	·
Potassium Permanganate, 10%	R	R	R	R	R	R	R	·	R	R	R	R	R	R	R	·	R	R	R	·	·	R	R	·	·
Potassium Permanganate, 25%	NR	NR	R	R	·	R	·	·	R	R	R	R	·	R	R	NR	R	R	·	·	·	R	R	·	·

Notes—Designations used are: R = Recommended; NR = Not Recommended; · = No data. All temperatures are in °F, and all chemicals are in solution in the pure or concentrated state unless indicated otherwise.

Chemical	PVC 70	PVC 140	CPVC 70	CPVC 140	CPVC 185	Polypropylene 70	Polypropylene 150	Polypropylene 180	PVDF 70	PVDF 150	PVDF 250	Polyethylene 70	Polyethylene 140	Fiberglass-Reinforced Epoxy 70	FRE 150	FRE 250	Fiberglass-Reinforced Vinyl/Polyester 70	FRVP 150	FRVP 250	Neoprene 70	Neoprene 185	EPDM 70	EPDM 185	Viton® 70	Viton® 185
Potassium Persulfate	R	R	R	R	·	R	R	·	R	R	·	R	R	R	R	·	R	R	R	·	·	R	R	·	·
Potassium Sulfate	R	R	R	R	R	R	R	R	R	R	R	R	R	R	R	R	R	R	R	R	R	R	R	R	R
Potassium Sulfide	R	R	R	R	·	R	R	R	R	R	R	R	R	·	·	·	·	·	·	·	·	·	·	·	·
Propane	R	R	R	R	·	R	·	·	R	R	R	R	R	R	·	·	·	·	·	R	NR	NR	NR	R	R
Propargyl Alcohol	R	NR	R	NR	NR	R	·	·	R	R	·	R	R	·	·	·	·	·	·	·	·	·	·	·	·
Propyl Alcohol	R	NR	R	NR	NR	R	·	·	R	R	·	R	R	·	·	·	·	·	·	R	R	R	R	R	R
Propylene Dichloride	NR	NR	NR	NR	NR	NR	NR	NR	·	·	·	NR	NR	·	·	·	·	·	·	·	·	·	·	·	·
Propylene Gycol	·	·	·	·	·	·	·	·	·	·	·	R	R	R	R	·	·	·	·	·	·	·	·	·	·
Propylene Oxide	NR	NR	·	·	·	·	·	·	NR	NR	NR	·	·	·	·	·	·	·	·	NR	NR	R	NR	NR	NR
Pyridine	NR	NR	NR	NR	NR	R	R	R	NR	NR	NR	·	·	NR	NR	NR	NR	NR	NR	NR	NR	R	NR	NR	NR
Pyrogallic Acid	R	NR	·	·	·	·	·	·	·	·	·	·	·	·	·	·	·	·	·	·	·	·	·	·	·
Rayon Coagulating Bath	R	R	R	R	R	R	·	·	R	·	·	R	R	·	·	·	·	·	·	·	·	·	·	·	·
Salicyclic Acid	R	R	·	·	·	·	·	·	R	R	·	·	·	R	R	·	R	·	·	R	R	R	R	R	R
Salicylaldehyde	·	·	·	·	·	·	·	·	R	R	NR	·	·	·	·	·	·	·	·	·	·	·	·	·	·
Selenic Acid	R	R	R	R	R	R	·	·	R	·	·	R	R	·	·	·	·	·	·	·	·	·	·	·	·
Silicic Acid	R	R	·	·	·	·	·	·	·	·	·	R	R	R	R	·	R	R	·	·	·	R	·	·	·
Silicone Oil	R	R	R	R	R	R	R	·	R	·	·	·	·	·	·	·	·	·	·	R	R	R	R	R	R
Silver Chloride	·	·	·	·	·	·	·	·	·	·	·	R	R	·	·	·	·	·	·	·	·	·	·	·	·
Silver Cyanide	R	R	R	R	R	R	R	R	R	R	R	R	R	R	·	·	R	R	·	·	·	R	R	·	·
Silver Nitrate	R	R	R	R	R	R	R	R	R	R	R	R	R	R	R	R	R	R	R	R	R	R	R	R	R
Silver Sulfate	R	R	·	·	·	·	·	·	·	·	·	·	·	·	·	·	·	·	·	·	·	R	R	R	R
Soaps	R	R	R	R	R	R	R	·	R	R	·	R	R	R	R	·	·	·	·	R	R	R	R	R	R
Soap Solutions	R	R	R	R	R	R	R	R	R	R	·	R	R	R	R	·	R	R	·	R	NR	R	R	R	R
Sodium Acetate	R	R	R	R	R	R	R	R	R	R	R	R	R	R	R	·	R	R	R	R	NR	R	R	NR	NR
Sodium Alum	R	R	·	·	·	·	·	·	·	·	·	·	·	·	·	·	·	·	·	·	·	R	R	·	·
Sodium Benzoate	R	R	R	R	R	R	R	R	R	R	R	R	R	R	R	·	R	R	R	·	·	R	R	·	·
Sodium Bicarbonate	R	R	R	R	R	R	R	R	R	R	R	R	R	R	R	·	R	R	·	R	R	R	R	R	R
Sodium Bichromate	R	R	·	·	·	·	·	·	·	·	·	·	·	·	·	·	·	·	·	·	·	R	R	·	·
Sodium Bisulfate	R	R	R	R	R	R	R	R	R	R	R	R	R	R	R	R	R	R	R	R	R	R	R	R	R
Sodium Bisulfite	R	R	R	R	R	R	R	R	R	R	R	R	·	R	R	·	R	R	R	R	R	R	R	R	R
Sodium Borate	R	R	R	R	R	R	·	·	R	R	R	R	R	·	·	·	·	·	·	R	R	R	R	R	R
Sodium Bromide	R	R	R	R	R	R	R	R	R	R	R	R	R	R	R	·	R	R	·	·	·	R	R	·	·
Sodium Carbonate	R	R	R	R	R	R	R	R	R	R	R	R	R	R	R	·	R	R	·	·	·	R	R	R	R
Sodium Chlorate	R	·	R	R	·	R	R	R	R	R	R	R	·	R	R	·	·	·	·	R	NR	R	R	R	R
Sodium Chloride	R	R	R	R	R	R	R	R	R	R	R	R	R	R	R	R	R	R	R	R	R	R	R	R	R
Sodium Chlorite	NR	NR	NR	NR	NR	R	R	·	R	·	·	·	·	R	R	·	R	·	·	·	·	R	R	·	·
Sodium Cyanide	R	R	R	R	R	R	R	R	R	R	R	R	R	R	R	·	R	R	·	R	R	R	R	R	R
Sodium Dichromate	R	R	R	R	R	R	R	·	R	R	·	R	R	R	R	·	R	R	·	·	·	·	·	·	·
Sodium Ferricyanide	R	R	R	R	R	R	R	·	R	R	R	R	R	R	R	·	R	R	·	·	·	·	·	·	·
Sodium Ferrocyanide	R	R	R	R	R	R	R	·	R	R	R	R	R	R	R	R	R	R	·	·	·	·	·	·	·
Sodium Fluoride	R	R	R	R	R	R	R	R	R	R	R	R	R	R	R	·	R	R	·	R	R	R	R	R	NR
Sodium Hydroxide, 15%	R	R	R	R	R	R	R	R	R	R	·	R	R	R	R	·	R	R	·	R	R	R	R	R	NR
Sodium Hydroxide, 30%	R	R	R	R	R	R	R	R	·	·	·	R	R	R	R	·	NR	NR	NR	R	R	R	R	R	NR
Sodium Hydroxide, 50%	R	R	R	R	R	R	R	R	R	R	NR	R	R	R	R	·	NR	NR	NR	R	NR	R	·	R	NR
Sodium Hydroxide, 70%	R	R	R	R	·	R	R	R	NR	·	·	R	R	R	R	·	NR	NR	NR	·	·	·	·	·	·
Sodium Hypochlorite	R	R	R	R	R	R	NR	NR	R	R	R	R	R	R	·	·	R	R	·	R	R	R	NR	R	R
Sodium Iodide	·	·	·	·	·	·	·	·	·	·	·	·	·	·	·	·	·	·	·	·	·	·	·	·	·
Sodium Metaphosphate	R	R	R	R	·	R	·	·	R	R	R	·	·	·	·	·	·	·	·	R	NR	R	R	R	R
Sodium Nitrate	R	R	R	R	R	R	R	R	R	R	R	R	R	R	R	·	R	R	R	R	NR	R	R	R	NR
Sodium Nitrite	R	R	R	R	R	R	R	R	R	R	R	R	R	R	R	·	R	R	·	·	·	R	R	R	R
Sodium Palmitrate Solution 5%	R	R	R	R	R	R	·	·	R	·	·	·	·	·	·	·	·	·	·	·	·	·	·	·	·
Sodium Perborate	R	R	R	R	R	R	R	R	R	·	·	·	·	·	·	·	·	·	·	R	NR	R	R	R	R
Sodium Perchlorate	R	R	·	·	·	·	·	·	·	·	·	·	·	·	·	·	·	·	·	·	·	·	·	·	·
Sodium Peroxide	R	R	R	R	R	R	R	R	R	R	·	·	·	R	·	NR	·	·	·	R	NR	R	R	R	R
Sodium Phosphate, Alkaline	R	R	R	R	R	R	R	R	R	R	·	·	·	R	R	R	·	·	·	R	NR	R	R	R	R
Sodium Phosphate Acid	R	R	·	·	·	R	R	R	·	·	·	R	R	R	R	·	·	·	·	R	NR	R	R	R	R
Sodium Phosphate, Neutral	R	R	R	R	R	R	R	R	R	R	R	R	R	R	R	R	R	R	·	R	·	R	R	R	R
Sodium Silicate	R	R	R	R	R	R	R	R	R	R	R	R	R	R	R	·	R	R	·	R	R	R	R	R	R
Sodium Sulfate	R	R	R	R	R	R	R	R	R	R	R	R	R	R	R	R	R	R	R	·	·	R	R	R	R
Sodium Sulfide	R	R	R	R	R	R	R	R	R	R	R	R	R	R	R	·	R	R	·	·	·	R	R	R	R
Sodium Sulfite	R	R	R	R	R	R	R	R	R	R	R	R	R	R	R	·	R	R	R	·	·	R	R	R	R
Sodium Thiosulfate	R	R	R	R	·	R	R	R	R	R	R	R	R	R	·	·	NR	NR	NR	R	R	R	R	R	R

Notes—Designations used are: R = Recommended; NR = Not Recommended; · = No data. All temperatures are in °F, and all chemicals are in solution in the pure or concentrated state unless indicated otherwise.

IN DEGREES FAHRENHEIT

Chemical	PVC		CPVC			Polypropylene			PVDF			Polyethylene		Fiberglass-Reinforced Epoxy			Fiberglass-Reinforced Vinyl/Polyester			Neoprene		EPDM		Viton®	
	70	140	70	140	185	70	150	180	70	150	250	70	140	70	150	250	70	150	250	70	185	70	185	70	185
Sour Crude Oil	R	R	R	R	R	R	·	·	R	R	R	·	·	·	·	·	·	·	·	NR	NR	NR	NR	NR	NR
Stannic Chloride	R	R	R	R	R	R	R	·	R	R	R	R	R	R	R	·	R	R	R	R	NR	R	R	R	R
Stannous Chloride	R	R	R	R	R	R	R	R	R	R	R	R	R	R	R	·	R	R	R	R	R	R	R	R	R
Starch	R	R	·	·	·	·	·	·	·	·	·	R	R	·	·	·	·	·	·	·	·	R	R	R	R
Stearic Acid	R	R	R	R	R	R	·	·	R	R	R	R	·	R	R	·	R	R	R	R	NR	R	NR	·	·
Stoddard's Solvent	NR	NR	NR	NR	NR	R	·	·	R	R	R	R	·	·	·	·	·	·	·	R	NR	NR	NR	R	·
Succinic Acid	R	R	R	·	·	R	R	·	R	R	·	·	·	·	·	·	·	·	·	·	·	·	·	·	·
Sulfamic Acid, 20%	R	R	·	·	·	R	R	R	·	·	·	·	·	R	R	·	·	·	·	·	·	·	·	·	·
Sulfonated Detergents	R	R	R	R	R	R	·	·	R	R	·	·	·	·	·	·	R	R	·	·	·	·	·	·	·
Sulfate Liquors	R	R	R	R	R	R	R	·	R	R	·	·	·	·	·	·	·	·	·	·	·	·	·	·	·
Sulfite Liquors	R	R	R	R	·	R	·	·	R	R	·	R	R	·	·	·	R	R	·	R	NR	R	NR	R	R
Sulfur	R	R	R	R	R	NR	NR	NR	R	R	R	R	R	R	R	·	·	·	·	R	R	R	R	R	R
Sulfur Chloride	R	R	R	R	R	NR	NR	NR	R	·	·	NR	NR	NR	NR	NR	R	·	·	NR	NR	NR	NR	R	R
Sulfur Dioxide, Dry	R	R	R	R	·	R	·	·	R	R	·	R	·	R	·	·	R	R	R	NR	NR	R	R	R	R
Sulfur Dioxide, Wet	NR	NR	R	R	·	R	·	·	R	R	·	NR	NR	R	·	·	R	R	·	R	NR	R	R	R	R
Sulfur Trioxide	R	R	R	R	R	NR	NR	NR	·	NR	NR	NR	NR	NR	NR	NR	R	R	·	NR	NR	R	NR	R	R
Sulfur Trioxide, Gas	R	R	·	·	·	·	·	·	·	·	·	·	·	·	·	·	·	·	·	·	·	·	·	R	R
Sulfuric Acid, 10%	R	R	R	R	R	R	R	R	R	R	R	R	R	R	R	NR	R	R	R	R	R	R	NR	R	R
Sulfuric Acid, 30%	R	R	R	R	R	R	R	R	R	R	R	R	R	R	R	NR	R	R	R	R	NR	R	NR	R	R
Sulfuric Acid, 50%	R	R	R	R	R	R	·	NR	R	R	R	R	·	R	R	NR	R	R	R	NR	NR	R	NR	R	R
Sulfuric Acid, 60%	R	R	R	R	R	R	R	·	R	R	R	R	·	R	R	NR	R	R	NR	NR	NR	NR	NR	R	R
Sulfuric Acid, 70%	R	R	R	R	R	·	NR	NR	R	R	R	R	·	R	NR	NR	R	R	NR	NR	NR	NR	NR	R	R
Sulfuric Acid, 80%	NR	NR	R	R	R	·	NR	NR	R	R	·	R	NR	NR	NR	NR	R	NR	NR	NR	NR	NR	NR	R	R
Sulfuric Acid, 90%	NR	NR	R	NR	NR	NR	NR	NR	R	R	·	R	NR	NR	NR	NR	NR	NR	NR	NR	NR	NR	NR	R	R
Sulfuric Acid, 93%	R	NR	R	NR	NR	R	NR	NR	R	R	·	R	NR	NR	NR	NR	NR	NR	NR	NR	NR	NR	NR	R	R
Sulfuric Acid, 94%	NR	NR	R	NR	NR	R	NR	NR	R	R	·	R	NR	NR	NR	NR	NR	NR	NR	NR	NR	NR	NR	R	R
Sulfuric Acid, 95%	NR	NR	R	NR	NR	R	NR	NR	R	R	·	R	NR	NR	NR	NR	NR	NR	NR	NR	NR	NR	NR	R	R
Sulfuric Acid, 96%	NR	NR	NR	NR	NR	R	NR	NR	R	R	·	R	NR	NR	NR	NR	NR	NR	NR	NR	NR	NR	NR	R	R
Sulfuric Acid, 98%	NR	NR	NR	NR	NR	NR	NR	NR	R	R	NR	R	NR	NR	NR	NR	R	NR	NR	NR	NR	NR	NR	R	R
Sulfuric Acid, 100%	NR	NR	NR	NR	NR	NR	NR	NR	R	R	·	NR	NR	NR	NR	NR	NR	NR	NR	NR	NR	NR	NR	R	R
Sulfurous Acid	R	R	R	R	R	R	R	·	R	R	R	·	·	R	R	NR	R	·	·	R	NR	R	NR	R	R
Tall Oil	R	R	R	R	R	R	R	R	R	R	R	R	·	·	·	·	·	·	·	·	·	·	·	R	R
Tannic Acid	R	R	R	R	R	R	R	R	R	R	R	R	R	R	R	·	R	R	·	R	R	R	R	R	R
Tanning Liquors	R	R	R	R	R	R	·	·	R	·	·	R	R	·	·	·	·	·	·	·	·	·	·	R	R
Tar	·	·	·	·	·	·	·	·	·	·	·	·	·	·	·	·	·	·	·	NR	NR	NR	NR	R	R
Tartaric Acid	R	R	R	R	R	R	R	·	R	R	R	R	R	R	R	·	R	R	R	R	NR	R	NR	R	R
Tetraethyl Lead	R	·	R	·	·	R	·	·	R	R	R	·	·	R	R	·	·	·	·	R	NR	NR	NR	R	R
Tetrahydrodurane	NR	NR	NR	NR	NR	NR	NR	NR	NR	NR	NR	·	·	·	·	·	·	·	·	·	·	NR	NR	·	·
Tetrahydrofuran	NR	NR	NR	NR	NR	NR	NR	NR	NR	NR	NR	NR	NR	·	·	·	·	·	·	NR	NR	NR	NR	NR	NR
Tetralin	NR	NR	NR	NR	NR	NR	NR	NR	R	R	·	·	·	·	·	·	·	·	·	NR	NR	NR	NR	R	R
Tetra Sodium Pyrophosphate	R	R	R	R	·	·	·	·	R	·	·	·	·	·	·	·	·	·	·	·	·	·	·	·	·
Thionyl Chloride	NR	NR	NR	NR	NR	R	NR	NR	NR	NR	NR	NR	NR	·	·	·	NR	NR	NR	NR	NR	NR	NR	R	NR
Thread Cutting Oils	R	R	R	R	R	R	·	·	R	R	·	·	·	·	·	·	·	·	·	·	·	NR	NR	·	·
Tin Chloride	R	R	R	R	R	R	NR	NR	·	·	·	·	·	·	·	·	·	·	·	·	·	·	·	·	·
Tirpineol	R	·	·	·	·	·	·	·	·	·	·	·	·	·	·	·	·	·	·	NR	NR	NR	NR	R	R
Titanium Tetrachloride	NR	NR	NR	NR	NR	NR	NR	NR	NR	NR	NR	NR	NR	·	·	·	·	·	·	NR	NR	NR	NR	R	R
Toluene Toluol	NR	NR	NR	NR	NR	NR	NR	NR	R	R	·	NR	NR	R	R	·	R	R	·	NR	NR	NR	NR	R	R
Tomato Juice	R	R	R	R	R	NR	NR	NR	R	R	·	R	R	R	R	·	·	·	·	·	·	·	·	R	R
Transformer Oil	R	R	R	R	R	R	NR	NR	R	·	·	R	·	R	R	NR	·	·	·	R	NR	NR	NR	R	R
Transformer Oil, DTE/30	R	R	R	R	R	R	R	R	R	R	·	R	R	·	·	·	·	·	·	R	NR	NR	NR	R	R
Tributyl Phosphate	NR	NR	NR	NR	NR	R	NR	NR	R	NR	NR	·	·	·	·	·	·	·	·	NR	NR	R	NR	NR	NR
Tributyl Citrate	R	·	·	·	·	·	·	·	·	·	·	·	·	·	·	·	·	·	·	·	·	·	·	·	·
Trichloroacetic Acid	R	·	R	·	·	R	·	·	R	NR	NR	·	·	R	R	·	R	R	·	NR	NR	R	NR	NR	NR
Trichloroethylene	NR	NR	NR	NR	NR	R	NR	NR	R	R	R	NR	NR	R	R	·	R	R	·	NR	NR	NR	NR	R	R
Triethanolamine	R	NR	NR	·	·	R	R	R	R	·	·	R	NR	R	R	·	R	·	·	·	·	R	NR	NR	NR
Triethylamine	R	R	R	R	·	NR	NR	NR	R	R	·	R	R	R	NR	NR	·	·	·	·	·	·	·	R	R
Trimethylpropane	R	·	R	·	·	NR	NR	NR	R	R	·	·	·	·	·	·	·	·	·	·	·	·	·	R	·
Trisodium Phosphate	R	R	R	R	R	R	R	·	R	R	R	R	R	R	R	·	R	R	R	·	·	·	·	R	R
Turpentine	NR	NR	R	·	·	NR	NR	NR	R	R	R	R	R	R	R	NR	R	·	·	NR	NR	NR	NR	R	R
Urea	R	R	R	R	R	R	R	R	R	R	R	R	R	R	R	·	·	·	·	·	·	R	R	R	R
Urine	R	R	R	R	R	R	R	R	R	R	R	R	R	·	·	·	·	·	·	·	·	R	R	R	R

Notes—Designations used are: R = Recommended; NR = Not Recommended; · = No data. All temperatures are in °F, and all chemicals are in solution in the pure or concentrated state unless indicated otherwise.

Chemical	PVC		CPVC			Polypropylene			PVDF			Polyethylene		Fiberglass-Reinforced Epoxy			Fiberglass-Reinforced Vinyl/Polyester			Neoprene		EPDM		Viton®	
IN DEGREES FAHRENHEIT	70	140	70	140	185	70	150	180	70	150	250	70	140	70	150	250	70	150	250	70	185	70	185	70	185
Vaseline	R	R	R	R	R	R	·	NR	R	R	R	·	·	·	·	·	·	·	·	·	·	NR	NR	·	·
Vegetable Oil	R	R	R	R	R	R	·	·	R	R	R	·	·	·	·	·	·	·	·	NR	NR	NR	NR	R	R
Vinegar	R	R	R	R	NR	R	R	·	R	R	R	R	R	R	R	·	R	·	·	R	NR	R	R	R	R
Vinegar, White	R	·	R	R	R	R	R	·	R	R	R	·	·	·	·	·	·	·	·	R	NR	R	R	R	R
Vinyl Acetate	NR	NR	NR	NR	NR	·	·	·	R	R	R	NR	NR	R	R	·	·	·	·	·	·	·	·	·	·
Water	R	R	R	R	R	R	R	R	R	R	R	R	R	R	R	R	R	R	R	R	R	R	R	R	R
Water, Acid Mine	R	R	R	R	R	R	R	·	R	R	R	R	R	·	·	·	·	·	·	·	·	·	·	R	R
Water, Demineralized	R	R	R	R	R	R	R	·	R	R	R	R	R	R	R	R	R	R	R	R	R	R	R	R	R
Water, Distilled or Fresh	R	R	R	R	R	R	R	R	R	R	R	R	R	R	R	R	R	R	·	·	·	R	R	R	R
Water, Potable	R	R	R	R	R	R	R	R	R	R	R	R	R	R	R	R	R	R	R	R	R	R	R	R	R
Water, Sale	R	R	R	R	R	R	R	R	R	R	R	R	R	R	R	R	·	·	·	R	R	R	R	R	R
Water, Sea	R	R	R	R	R	R	R	R	R	R	R	·	·	R	R	R	R	R	R	R	R	R	R	R	R
Water, Sewage	R	R	R	R	R	R	R	R	R	R	R	·	·	·	·	·	·	·	·	·	·	·	·	R	R
Whiskey	R	R	R	R	R	R	R	·	R	R	R	R	R	·	·	·	·	·	·	R	R	R	R	R	R
White Liquor	R	R	R	R	R	R	·	·	R	R	R	·	·	·	·	·	·	·	·	·	·	·	·	R	R
Wines	R	R	R	R	R	R	R	·	R	R	R	R	R	·	·	·	·	·	·	R	R	R	R	R	R
Xylene (Xylol)	NR	NR	NR	NR	NR	NR	NR	NR	R	R	·	NR	NR	R	R	·	R	·	·	NR	NR	NR	NR	R	R
Zinc Chloride	R	R	R	R	R	R	R	R	R	R	R	R	R	R	R	R	R	R	R	R	R	R	R	R	R
Zinc Nitrate	R	R	R	R	R	R	R	R	R	R	R	R	R	R	R	R	R	R	R	R	R	R	R	R	R
Zinc Sulfate	R	R	R	R	R	R	R	R	R	R	R	R	R	R	R	R	R	R	R	R	R	R	R	R	R

E

Appendix

PIPE CAPACITY AND FRICTION-LOSS TABLES, AND MISCELLANEOUS DATA

Approximate Friction Loss in Thermoplastic Pipe Fittings in Equivalent Feet of Pipe

Nominal Pipe Size, in.	3/8	1/2	3/4	1	1 1/4	1 1/2	2	2 1/2	3	3 1/2	4	5	6	8
Tee, Side Outlet	3	4	5	6	7	8	12	15	16	20	22	28	32	38
90° Ell	1 1/2	1 1/2	2	2 3/4	4	4	6	8	8	10	12	14	18	22
45° Ell	3/4	3/4	1	1 3/8	1 3/4	2	2 1/2	3	4	4 1/2	5	6	8	10
Insert Coupling	—	1/2	3/4	1	1 1/4	1 1/2	2	3	3	—	4	—	6 1/4	—
Male-Female Insert Adapters	—	1	1 1/2	2	2 3/4	3 1/2	4 1/2	—	6 1/2	—	9	—	14	—

Friction Loss in Schedule 40 Plastic Pipe Velocity Measured in ft/sec; Loss in Feet of Water Head per 100 ft of Pipe.

Note: In the lower part of the table the columns are re-used for larger pipe sizes: the 1/2" column becomes 5" PIPE (from 20 gpm), the 3/4" column becomes 6" PIPE (from 40 gpm), the 1" column becomes 8" PIPE (from 120 gpm), and the 1 1/4" column becomes 10" PIPE (from 190 gpm).

Gal per Min	1/2" Vel	1/2" Loss	3/4" Vel	3/4" Loss	1" Vel	1" Loss	1 1/4" Vel	1 1/4" Loss	1 1/2" Vel	1 1/2" Loss	2" Vel	2" Loss	2 1/2" Vel	2 1/2" Loss	3" Vel	3" Loss	3 1/2" Vel	3 1/2" Loss	4" Vel	4" Loss
2	2.10	3.47	1.20	0.89																
4	4.23	12.7	2.41	3.29	1.49	1.01	.86	.27	.63	.12										
6	6.34	26.8	3.61	6.91	2.23	2.14	1.29	.57	.94	.26	.57	.09								
8	8.45	46.1	4.82	11.8	2.98	3.68	1.72	.95	1.26	.45	.77	.16	.52	.05						
10	10.6	69.1	6.02	17.9	3.72	5.50	2.14	1.44	1.57	.67	.96	.24	.65	.08	.43	.03				
12			7.22	24.9	4.46	7.71	2.57	2.02	1.89	.94	1.15	.37	.78	.11	.52	.05				
15			9.02	37.6	5.60	11.8	3.21	3.05	2.36	1.41	1.50	.51	.98	.17	.65	.07	.49	.03		
18			10.8	50.9	6.69	16.5	3.86	4.28	2.83	1.99	1.72	.70	1.18	.24	.78	.10	.58	.04		
20	5" PIPE		12.0	63.9	7.44	19.7	4.29	5.21	3.15	2.44	1.91	.86	1.31	.29	.87	.12	.65	.05	.51	.03
25					9.30	30.1	5.36	7.80	3.80	3.43	2.50	1.28	1.63	.43	1.09	.18	.81	.08	.64	.04
30	.49	.02			11.15	41.8	6.43	10.8	4.72	5.17	2.89	1.80	1.96	.61	1.30	.25	.97	.11	.77	.06
35	.57	.03			13.02	55.9	7.51	14.7	5.51	6.91	3.35	2.40	2.35	.81	1.52	.33	1.14	.15	.89	.08
40	.65	.04	6" PIPE		14.88	71.4	8.58	18.8	6.30	8.83	3.82	3.10	2.68	1.03	1.74	.43	1.30	.19	1.02	.10
45	.73	.04			16.70		9.65	23.5	7.08	10.9	4.30	3.85	3.02	1.32	1.95	.54	1.46	.24	1.15	.13
50	.82	.05	.57	.02			10.72	28.2	7.87	13.3	4.78	4.65	3.35	1.56	2.17	.65	1.62	.29	1.28	.16
55	.90	.06	.62	.02			11.78	33.8	8.66	16.0	5.26	5.55	3.69	1.88	2.39	.74	1.70	.34	1.41	.19
60	.98	.07	.68	.03			12.87	40.0	9.44	18.6	5.74	6.53	4.02	2.19	2.60	.90	1.95	.40	1.53	.22
65	1.06	.09	.74	.04			13.92	46.7	10.23	21.6	6.21	7.56	4.36	2.53	2.82	1.02	2.00	.47	1.66	.25
70	1.14	.10	.79	.04			15.01	53.1	11.02	24.9	6.69	8.64	4.69	2.91	3.04	1.21	2.27	.54	1.79	.30
75	1.22	.11	.85	.05			16.06	60.6	11.80	28.2	7.17	9.82	5.03	3.33	3.25	1.41	2.32	.60	1.91	.34
80	1.31	.13	.91	.05			17.16	68.2	12.69	32.0	7.65	11.1	5.36	3.71	3.49	1.54	2.60	.69	2.04	.38
85	1.39	.15	.96	.06			18.21	77.0	13.38	35.3	8.13	12.5	5.70	3.81	3.69	1.66	2.62	.76	2.17	.42
90	1.47	.16	1.02	.07					14.71	39.5	8.61	13.8	6.03	4.61	3.91	1.92	2.92	.85	2.30	.47
95	1.55	.18	1.08	.07					14.95	43.7	9.08	15.3	6.37	5.07	4.12	2.04	2.93	.96	2.42	.53
100	1.63	.19	1.13	.08					15.74	47.9	9.56	16.8	6.70	5.64	4.34	2.33	3.25	1.03	2.55	.57
110	1.79	.23	1.25	.10					17.31	57.3	10.5	20.2	7.37	6.81	4.77	2.82	3.57	1.25	2.81	.69
120	1.96	.27	1.36	.11	8" PIPE				18.89	67.2	11.5	23.5	8.04	7.89	5.21	3.29	3.99	1.45	3.06	.80
130	2.12	.31	1.47	.13					20.46	78.0	12.4	27.3	8.71	8.79	5.64	3.81	4.22	1.68	3.31	.93
140	2.29	.36	1.59	.15	.90	.04			22.04	89.3	13.4	31.5	9.38	10.5	6.08	4.32	4.54	1.93	3.57	1.07
150	2.45	.41	1.70	.17	.96	.04			23.6		14.3	35.7	10.00	12.0	6.51	4.93	4.87	2.19	3.82	1.23
160	2.61	.46	1.80	.19	1.02	.05					15.3	40.4	10.7	13.6	6.94	5.54	5.19	2.47	4.08	1.37
170	2.77	.51	1.92	.21	1.08	.05					16.3	45.1	11.4	16.0	7.36	6.25	5.52	2.75	4.33	1.53
180	2.94	.57	2.04	.24	1.15	.06					17.2	50.3	12.1	16.8	7.81	6.58	5.85	3.07	4.60	1.70
190	3.10	.63	2.16	.26	1.21	.07	10" PIPE				18.2	55.5	12.7	18.6	8.24	7.28	6.17	3.39	4.84	1.88
200	3.27	.70	2.27	.29	1.28	.07					19.1	60.6	13.4	20.3	8.68	8.36	6.50	3.73	5.11	2.06
220	3.59	.83	2.44	.34	1.40	.08	.90	.03			21.0	72.4	14.7	24.9	9.55	10.0	7.14	4.45	5.62	2.44
240	3.92	.98	2.67	.41	1.53	.10	.98	.03			22.9	85.5	16.1	28.7	10.4	11.8	7.79	5.22	6.13	2.91
260	4.25	1.13	2.89	.47	1.66	.12	1.06	.04			24.9	99.2	17.4	33.0	11.3	13.7	8.44	6.07	6.64	3.28
280	4.50	1.30	3.11	.54	1.79	.13	1.15	.04					18.8	38.1	12.2	15.7	9.09	6.95	7.15	3.85
300	4.90	1.48	3.33	.62	1.91	.15	1.22	.05					20.1	43.2	13.0	17.9	9.74	7.90	7.66	4.37
320	5.13	1.66	3.56	.69	2.05	.17	1.31	.06					21.6	48.4	13.9	20.1	10.40	8.88	8.17	4.93

(continued)

Friction Loss in Schedule 40 Plastic Pipe Velocity Measured in ft/sec; Loss in Feet of Water Head per 100 ft of Pipe.

Gal per Min	1/2" Vel	Loss	3/4" Vel	Loss	1" Vel	Loss	1 1/4" Vel	Loss	1 1/2" Vel	Loss	2" Vel	Loss	2 1/2" Vel	Loss	3" Vel	Loss	3 1/2" Vel	Loss	4" Vel	Loss
340	5.44	1.87	3.78	.76	2.18	.19	1.39	.07	12" PIPE				22.9	54.5	14.8	22.5	11.00	9.96	8.58	5.50
360	5.77	2.07	4.00	.86	2.30	.21	1.47	.07					24.2	60.2	15.6	24.9	11.70	11.0	9.10	6.15
380	6.19	2.28	4.22	.94	2.43	.24	1.55	.08	1.08	.03			25.6	66.7	16.5	27.7	12.3	12.2	9.59	6.58
400	6.44	2.5	4.43	1.03	2.60	.25	1.63	.09	1.14	.04			26.8	73.3	17.4	30.6	13.0	13.4	10.10	7.52
450	7.20	3.1	5.00	1.29	2.92	.32	1.84	.11	1.28	.05					19.5	36.7	13.9	16.7	11.49	9.31
500	8.02	3.8	5.56	1.36	3.19	.39	2.04	.13	1.42	.05					21.7	46.1	16.2	20.3	12.6	11.3
550	8.82	4.5	6.11	1.86	3.52	.46	2.24	.16	1.56	.06					23.9	55.0	17.9	24.3	13.0	13.5
600	9.62	5.3	6.65	2.19	3.85	.54	2.45	.18	1.70	.07					26.0	64.4	19.5	28.5	15.10	15.8
650	10.40	6.2	7.22	2.53	4.16	.63	2.65	.21	1.84	.09					28.2		21.1	33.0	16.40	18.3
700	11.2	7.1	7.78	2.92	4.46	.72	2.86	.24	1.99	.10							22.7	37.9	17.60	21.1
750	12.0	8.1	8.34	3.35	4.80	.82	3.06	.28	2.13	.11							24.4	43.0	18.90	24.0
800	12.8	9.1	8.90	3.74	5.10	.89	3.26	.31	2.27	.13							26.0	48.4	20.20	26.8
850	13.6	10.2	9.45	4.21	5.48	1.03	3.47	.35	2.41	.15							27.6	54.1	21.4	30.1
900	14.4	11.3	10.0	4.75	5.75	1.16	3.67	.39	2.56	.16									22.7	33.4
950	15.2	12.5	10.5	5.26	6.06	1.35	3.88	.43	2.70	.18										
1000	16.0	13.7	11.1	5.66	6.38	1.40	4.08	.48	2.84	.19										
1100	17.6	16.4	12.2	6.84	7.03	1.65	4.49	.56	3.13	.23										
1200	19.61	19.2	13.3	8.04	7.66	1.96	4.90	.66	3.41	.27										
1300	20.8		14.4	8.6	8.30	2.28	5.31	.76	3.69	.31										
1400	22.4		15.6	10.6	8.95	2.59	5.71	.88	3.98	.37										
1500	24.0		16.7	12.0	9.58	2.93	6.12	1.00	4.26	.42										
1600	25.6		17.8	12.6	10.21	3.29	6.53	1.12	4.55	.46										
1800			20.0		11.50	4.13	7.35	1.39	5.11	.57										
2000			22.2		12.78	5.03	8.16	1.69	5.68	.70										
2200			24.4		14.05	6.00	8.98	1.99	6.25	.85										
2400			26.7		15.32	6.7	9.80	2.37	6.81	.98										
2600							10.61	2.73	7.38	1.14										
2800							11.41	3.15	7.95	1.29										
3000							12.24	3.58	8.52	1.48										
3200							13.05	3.7	9.10	1.65										
3500							14.30	4.74	9.95	1.96										
3800							15.51	6.3	10.80	2.30										
4200									11.92	2.76										
4500									12.78	3.24										
5000									14.20	3.95										

*Data shown are calculated from Williams and Hazen formula $H = (3.023/C^{1.852})(V^{1.852}/D^{1.167})$ using $C = 150$. For water at 60° F. Where H = head loss, V = fluid velocity ft/sec, D = diameter of pipe, ft. and C = coefficient representing roughness of pipe interior surface.

Friction Loss in Schedule 80 Plastic Pipe Velocity Measured in ft./sec., Loss in Feet of Water Head per 100 ft. of Pipe.

Gal per Min	1/2" Vel	Loss	3/4" Vel	Loss	1" Vel	Loss	1 1/4" Vel	Loss	1 1/2" Vel	Loss	2" Vel	Loss	2 1/2" Vel	Loss	3" Vel	Loss	3 1/2" Vel	Loss	4" Vel	Loss
2	2.74	6.72	1.48	1.51																
4	5.48	24.2	2.97	5.45	1.79	1.54	1.00	.39	.73	.177										
6	8.23	51.2	4.45	11.5	2.68	3.34	1.50	.82	1.09	.375	.65	.107								
8	11.0	86.9	5.94	19.6	3.57	5.69	2.00	1.39	1.45	.64	.87	.183	.61	.077						
10	13.7	132.0	7.42	29.6	4.46	8.60	2.50	2.10	1.82	.96	1.09	.276	.76	.115	.485	.039				
12			8.91	41.5	5.36	12.0	3.00	2.94	2.18	1.35	1.30	.387	.91	.161	.572	.055				
15			11.1	62.7	6.7	22.9	3.76	4.45	2.72	2.04	1.63	.585	1.14	.243	.727	.083	.54	.035		
18			13.4	87.9	8.03	25.5	4.50	6.25	3.27	2.86	1.96	.818	1.36	.340	.873	.116	.65	.056		
20	5" PIPE		14.8	107	8.92	30.9	5.00	7.57	3.63	3.47	2.17	.996	1.51	.414	.97	.140	.72	.068	.56	.037
25					11.2	58.8	6.25	11.4	4.55	5.25	2.71	1.51	1.9	.625	1.21	.212	.90	.103	.695	.055
30	.53	.025			13.4	65.3	7.50	16.0	5.45	7.38	3.26	2.11	2.27	.874	1.44	.297	1.08	.145	.84	.077
35	.62	.034			15.6	86.9	8.75	21.3	6.38	9.78	3.80	2.81	2.65	1.16	1.70	.396	1.26	.192	.973	.103
40	.71	.043	6" PIPE		17.9	111	10.0	27.3	7.26	12.5	4.35	3.59	3.03	1.49	1.94	.507	1.44	.246	1.12	.132
45	.795	.054					11.2	33.9	8.26	15.6	4.89	4.46	3.41	1.86	2.18	.629	1.63	.306	1.25	.164
50	.88	.065	.62	.027			12.5	41.3	9.08	18.9	5.43	5.41	3.79	2.25	2.42	.766	1.80	.372	1.40	.199
55	.973	.078	.676	.032			13.7	49.2	10.00	32.0	5.98	6.44	4.16	2.68	2.67	.912	1.99	.443	1.53	.237
60	1.06	.091	.74	.039			15.0	57.8	10.9	26.5	6.52	7.61	4.54	3.16	2.92	1.07	2.17	.522	1.67	.279
65	1.15	.106	.80	.044			16.1	67.0	11.8	30.7	7.06	8.84	4.92	3.66	3.14	1.25	2.35	.604	1.81	.323
70	1.23	.121	.86	.051			17.5	77.1	12.7	35.3	7.61	10.1	5.30	4.20	3.39	1.43	2.53	.691	1.95	.371
75	1.33	.138	.923	.057			18.8	87.4	13.6	40.1	8.15	11.5	5.68	4.79	3.64	1.62	2.70	.787	2.08	.421
80	1.41	.155	.98	.065			20.0	98.2	14.5	45.2	8.69	12.9	6.05	5.36	3.88	1.83	2.89	.888	2.23	.475
85	1.50	.174	1.04	.072			21.2	110	15.4	50.3	9.03	14.5	6.43	6.02	4.10	2.04	3.05	.992	2.34	.531
90	1.59	.193	1.11	.080			22.5	122	16.3	55.9	9.78	16.1	6.81	6.53	4.33	2.27	3.25	1.10	2.51	.592
95	1.67	.213	1.20	.089					17.2	62.0	10.3	17.8	7.19	7.38	4.57	2.51	3.42	1.21	2.64	.652
100	1.76	.234	1.23	.098					18.2	68.2	10.9	19.6	7.57	8.13	4.85	2.76	3.67	1.34	2.79	.719
110	1.95	.279	1.36	.117					20.0	81.3	12.0	23.4	8.33	9.68	5.33	3.29	3.97	1.60	3.07	.855
120	2.11	.329	1.48	.137	8" PIPE				21.8	95.4	13.0	27.4	9.08	11.4	5.80	3.87	4.33	1.88	3.35	1.00
130	2.3	.381	1.60	.159					23.6	111	14.1	31.8	9.84	13.2	6.30	4.48	4.69	2.18	3.63	1.16
140	2.47	.437	1.72	.182	.98	.047			25.4	127	15.2	36.5	10.6	15.1	6.80	5.12	5.05	2.50	3.91	1.33
150	2.65	.496	1.85	.207	1.05	.054					16.3	41.5	11.3	17.2	7.27	5.87	5.41	2.84	4.19	1.52
160	2.82	.559	1.97	.234	1.12	.059					17.4	46.7	12.1	19.4	7.75	6.58	5.78	3.20	4.47	1.71
170	3.0	.626	2.08	.261	1.19	.067					18.5	52.2	12.9	21.7	8.20	7.37	6.14	3.58	4.75	1.91
180	3.16	.696	2.22	.290	1.26	.074					19.6	58.3	13.6	24.1	8.60	8.18	6.50	3.97	5.02	2.12
190	3.36	.769	2.34	.321	1.33	.082					20.6	64.4	14.4	26.6	9.20	9.05	6.85	4.39	5.30	2.35

(continued)

Friction Loss in Schedule 80 Plastic Pipe Velocity Measured in ft./sec., Loss in Feet of Water Head per 100 ft. of Pipe.

Gal per Min	1/2" Vel	1/2" Loss	3/4" Vel	3/4" Loss	1" Vel	1" Loss	1 1/4" Vel	1 1/4" Loss	1 1/2" Vel	1 1/2" Loss	2" Vel	2" Loss	2 1/2" Vel	2 1/2" Loss	3" Vel	3" Loss	3 1/2" Vel	3 1/2" Loss	4" Vel	4" Loss
200	3.52	.846	2.46	.353	1.41	.090					21.7	70.5	15.1	29.3	9.70	9.96	7.22	4.84	5.58	2.58
220	3.88	1.01	2.71	.421	1.55	.108					23.9	84.1	16.7	34.9	10.6	11.9	7.94	5.78	6.14	3.08
240	4.23	1.18	2.96	.484	1.69	.126					26.1	98.7	18.2	41.0	11.6	13.9	8.66	6.77	6.70	3.62
260	4.58	1.37	3.20	.573	1.83	.147					28.3	115	19.7	47.5	12.6	16.2	9.38	7.85	7.26	4.19
280	4.94	1.57	3.45	.658	1.97	.168							21.2	54.5	13.5	18.6	10.1	9.02	7.82	4.79
300	5.29	1.79	3.69	.747	2.11	.191							22.7	62.0	14.4	21.1	10.8	10.2	8.38	5.45
320	5.64	2.01	3.94	.841	2.24	.215							24.2	69.9	15.5	23.7	11.5	11.5	8.94	6.16
340	5.99	2.26	4.19	.940	2.39	.240							25.8	78.2	16.3	26.6	12.3	12.9	9.50	6.91
360	6.35	2.51	4.43	1.05	2.64	.261							27.2	86.9	17.4	29.5	13.0	14.3	10.0	7.66
380	6.70	2.77	4.68	1.16	2.68	.295							28.8	96.1	18.6	32.6	13.7	15.8	10.6	8.46
400	7.05	3.05	4.93	1.27	2.81	.325							30.3	106	19.4	35.9	14.4	17.4	11.2	9.31
450	7.95	3.79	5.54	1.58	3.16	.404									21.8	44.6	16.2	21.6	12.5	11.6
500	8.82	4.61	6.16	1.92	3.51	.493									23.2	54.1	18.1	26.3	14.0	14.1
550	9.70	5.50	6.77	2.29	3.86	.587									26.5	64.9	19.9	31.4	15.3	16.8
600	10.6	6.44	7.39	2.69	4.22	.686									29.1	76.1	21.7	36.9	16.7	19.7
650	11.5	7.47	8.00	3.12	4.57	.799											23.5	42.8	18.1	22.9
700	12.3	8.60	8.63	3.58	4.92	.916											25.3	48.9	19.5	26.2
750	13.2	9.77	9.24	4.07	5.27	1.04											27.1	55.9	20.9	29.8
800	14.1	11.0	9.85	4.58	5.62	1.17											28.9	61.6	22.3	33.6
850	15.0	12.3	10.5	5.12	5.97	1.31											30.7	70.5	23.7	37.6
900	15.9	13.7	11.1	5.69	6.32	1.46													25.1	41.8
950	16.7	15.1	11.7	6.29	6.67	1.61														
1000	17.6	16.6	12.3	6.91	7.03	1.77														
1100	19.4	19.8	13.5	8.27	7.83	2.11														
1200	21.1	23.3	14.8	9.73	8.43	2.48														
1300					9.13	2.87														
1400					9.83	3.30														
1500					10.5	3.75														
1600					11.2	4.23														
1800					12.6	5.26														
2000					14.1	6.39														
2200					15.5	7.80														
2400					16.9	8.93														
2600																				
2800																				
3000																				
3200																				
3500																				

*Data shown are calculated from Williams and Hazen formula $H = (3.023/C^{1.852})(V^{1.852}/D^{1.167})$ using $C = 150$. For water at 60° F. Where H = head loss, V = fluid velocity ft/sec, D = diameter of pipe, ft. and C = coefficient representing roughness of pipe interior surface.

Pressure Drop for PE Pipe, in psi per 100 ft PE Pipe with Standard ID (CS-255-63 and ASTM D 2239-67)

U.S. GPM	1/2 0.662	3/4 0.824	1 1.049	1 1/4 1.380	1 1/2 1.610	2 2.067	2 1/2 2.469	3 3.068	4 4.026	U.S. GPM
1	0.49									1
2	1.8	0.45	0.14							2
3	3.7	0.95	0.30							3
4	6.4	1.6	0.50	0.13	0.07					4
5	9.6	2.5	0.75	0.20	0.10					5
6	13.4	3.4	1.1	0.28	0.13					6
7		4.6	1.4	0.37	0.17					7
8		5.8	1.8	0.48	0.23	0.07				8
9		7.3	2.2	0.59	0.28	0.08				9
10		8.8	2.7	0.72	0.34	0.10				10
11		10.5	3.3	0.85	0.41	0.12				11
12		12.4	3.8	1.0	0.48	0.14	0.06			12
13			4.4	1.2	0.55	0.16	0.07			13
14			5.1	1.3	0.63	0.19	0.08			14
15			5.8	1.5	0.72	0.22	0.09			15
16			6.5	1.7	0.81	0.24	0.10			16
17			7.3	1.9	0.91	0.27	0.11			17
18			8.1	2.1	1.0	0.30	0.12			18
19			9.0	2.4	1.1	0.33	0.13			19
20			9.8	2.6	1.2	0.36	0.15	0.06		20
22			11.8	3.1	1.5	0.44	0.18	0.07		22
24			13.8	3.6	1.8	0.52	0.22	0.08		24
25			14.9	3.9	1.9	0.56	0.23	0.08		25
26				4.3	2.0	0.59	0.25	0.09		26

(continued)

Pressure Drop for PE Pipe, in psi per 100 ft PE Pipe with Standard ID (CS-255-63 and ASTM D 2239-67)

U.S. GPM	1/2 0.662	3/4 0.824	1 1.049	1¼ 1.380	1½ 1.610	2 2.067	2½ 2.469	3 3.068	4 4.026	U.S. GPM
28				4.8	2.3	0.68	0.29	0.10		28
30				5.5	2.6	0.77	0.32	0.11		30
35				7.3	3.5	1.0	0.43	0.15		35
40				9.3	4.4	1.3	0.54	0.19	0.06	40
45				11.6	5.5	1.6	0.69	0.24	0.07	45
50				14.1	6.6	2.0	0.83	0.29	0.08	50
60					9.3	2.8	1.2	0.40	0.11	60
70					12.4	3.7	1.6	0.54	0.15	70
80						4.7	2.0	0.69	0.19	80
90						5.9	2.5	0.86	0.23	90
100						7.1	3.0	1.0	0.28	100
120						10.1	4.2	1.5	0.39	120
140						13.3	5.6	1.9	0.52	140
160							7.2	2.5	0.66	160
180							8.9	3.1	0.83	180
200							10.8	3.8	1.0	200
220								4.5	1.2	220
240								5.3	1.4	240
260								6.1	1.6	260
280								7.0	1.9	280
300								8.0	2.1	300

Note: Values below solid lines are at velocities over 5 feet per second and should be selected with caution.

Approximate Friction Loss in Fittings in Equivalent Feet of Pipe

	½	¾	1	1¼	1½	2	2½	3	4
Straight 'L'	1.7	2.1	2.6	3.5	4.1	5.2	6.2	7.7	10.2
Long Sweep 'L'	0.47	0.62	0.79	1.03	1.21	1.55	1.85	2.3	3.0
Tee (Running)	1.1	1.4	1.8	2.3	2.7	3.5	4.2	5.2	6.8
Tee (Branch)	3.3	4.2	5.3	7.0	8.1	10.4	12.4	15.5	20.3

GPM = Gallons per minute

Carrying Capacity and Friction Loss for Thermoplastic Tubing, 0.090 in. Wall

Gallons per Minute	½ in. Velocity Feet per Second	Friction Head Feet	Friction Loss Pounds per Square Inch	¾ in. Velocity Feet per Second	Friction Head Feet	Friction Loss Pounds per Square Inch	1 in. Velocity Feet per Second	Friction Head Feet	Friction Loss Pounds per Square Inch	1¼ in. Velocity Feet per Second	Friction Head Feet	Friction Loss Pounds per Square Inch	1¾ in. Velocity Feet per Second	Friction Head Feet	Friction Loss Pounds per Square Inch
1	2.07	9.07	3.93	0.85	1.04	0.45									
2	4.13	18.14	7.85	1.69	2.07	0.90	0.92	0.84	0.36	0.57	0.27	0.12	0.28	0.07	0.030
5	10.32	102.13	44.22	4.23	11.65	5.04	2.29	2.61	1.13	1.43	0.83	0.31	0.71	0.18	0.075
7				5.93	21.39	9.26	3.21	4.79	2.07	2.01	1.53	0.66	1.00	0.28	0.12
10				8.46	40.74	17.64	4.58	9.12	3.95	2.86	2.91	1.26	1.42	0.53	0.23
15				12.69	86.33	37.38	6.87	19.33	8.37	4.29	6.16	2.67	2.13	1.12	0.48
20							9.15	32.93	14.26	5.72	10.50	4.55	2.85	1.91	0.83
25							11.44	49.78	21.55	7.16	15.87	6.87	3.56	2.89	1.25
30							13.73	69.77	30.21	8.59	22.24	9.63	4.27	4.05	1.76
35										10.02	29.59	12.81	4.98	5.39	2.33
40										11.45	37.89	16.41	5.69	6.91	2.99
45										12.88	47.13	20.41	6.40	8.59	3.72
50										14.31	57.28	24.80	7.11	10.44	4.52
60													8.54	14.63	6.33
70													9.96	19.47	8.43
75													10.67	22.12	9.58
80													11.38	24.93	10.79
90													12.80	31.00	13.42
100													14.23	37.68	16.32

(Independent variables: Gallons per minute and nominal pipe size OD. Dependent Variables: Velocity, Friction Head, and Pressure Drop per 100 Feet of Pipe, Interior Smooth.)

Earth Loads in Trench Conditions (lb/linear ft) $W_c = C_d w B_d B_c$

Depth of Cover	Type of Soil	4" Pipe Width of Trench (ft.)				6" Pipe Width of Trench (ft.)				8" Pipe Width of Trench (ft.)				10" Pipe Width of Trench (ft.)				12" Pipe Width of Trench (ft.)				15" Pipe Width of Trench (ft.)			
		0.75	1.00	1.25	1.5	1.0	1.5	2.0	2.5	1.5	2.0	2.5	3.0	2.0	2.5	3.0	3.5	2.0	2.5	3.0	3.5	2.0	2.5	3.0	3.5
3	Granular w/o Cohesion	54	63	68	74	94	110	120	123	143	156	160	167	196	200	209	214	235	240	251	257	293	300	314	321
	Sand and Gravel	65	73	79	84	109	125	136	140	163	177	181	191	221	227	238	242	265	272	286	291	331	340	358	363
	Sat. Top Soil	75	81	86	91	120	135	149	166	176	190	196	204	240	244	255	260	287	293	306	312	361	367	383	390
	Dry Clay	82	86	95	100	132	149	158	176	193	206	212	220	257	265	275	278	308	318	335	338	386	398	409	423
	Sat. Clay	95	100	106	112	150	166	177	184	216	230	239	244	287	298	305	309	345	358	366	371	431	448	457	464
3.5	Granular w/o Cohesion	57	68	75	79	101	118	131	140	153	170	182	184	213	227	230	244	255	273	275	293	319	341	344	366
	Sand and Gravel	70	79	87	93	118	138	150	158	180	194	206	211	243	257	264	278	292	309	316	334	365	386	396	417
	Sat. Top Soil	77	89	96	102	132	152	163	184	197	213	223	228	266	279	284	297	319	334	341	357	399	418	427	446
	Dry Clay	86	97	105	109	144	169	176	193	212	228	241	245	286	301	316	321	343	361	376	386	428	451	469	482
	Sat. Clay	99	110	119	123	163	184	196	207	239	255	270	278	318	337	348	355	382	404	418	427	477	507	522	534
4	Granular w/o Cohesion	50	72	81	81	107	129	146	157	168	190	204	208	238	255	255	262	286	306	310	314	357	383	387	393
	Sand and Gravel	72	86	95	101	129	151	167	178	196	212	232	238	271	290	297	301	325	348	357	361	407	435	446	452
	Sat. Top Soil	81	97	106	112	144	167	184	204	217	235	250	258	293	313	323	325	352	375	387	390	440	469	484	498
	Dry Clay	90	105	116	123	156	184	198	217	239	258	269	279	322	337	349	360	387	404	425	428	483	505	519	536
	Sat. Clay	106	121	131	137	181	204	222	231	265	288	301	310	360	376	385	397	432	451	461	487	540	564	577	609
6	Granular w/o Cohesion	66	82	97	108	123	161	188	203	209	245	264	286	306	329	351	372	367	395	428	446	459	494	536	558
	Sand and Gravel	82	102	116	130	151	193	219	234	244	277	295	316	340	364	389	400	426	457	488	511	533	572	610	638
	Sat. Top Soil	92	114	131	144	170	216	236	271	274	306	322	338	378	402	426	444	469	504	528	554	587	630	660	693
	Dry Clay	109	127	145	158	190	236	252	282	302	334	355	370	404	434	452	478	514	551	594	600	643	689	725	750
	Sat. Clay	130	155	168	182	231	271	299	313	340	386	394	405	455	483	500	553	583	610	648	659	729	762	811	824
8	Granular w/o Cohesion	68	88	105	120	131	178	214	242	232	279	315	337	349	393	421	446	418	472	505	536	523	590	612	669
	Sand and Gravel	85	109	130	145	163	216	258	281	288	337	370	393	414	455	475	507	503	547	589	636	628	684	736	795
	Sat. Top Soil	98	123	146	164	184	244	273	323	322	374	406	430	460	503	533	553	561	616	651	698	701	770	814	872
	Dry Clay	117	145	166	183	217	273	314	345	358	408	443	472	504	552	573	602	612	673	734	763	765	842	895	926
	Sat. Clay	141	173	200	217	258	323	361	391	441	461	503	523	572	620	643	677	705	762	796	849	882	953	995	1062
10	Granular w/o Cohesion	68	90	111	126	133	188	234	268	245	305	349	377	381	436	472	515	457	523	566	618	564	654	708	772
	Sand and Soil	87	114	137	157	169	234	283	322	314	376	424	460	465	522	572	600	552	628	673	718	690	785	842	898
	Sat. Top Soil	101	131	156	176	195	262	301	362	350	420	473	504	518	583	630	656	617	701	746	801	771	876	933	1001
	Dry Clay	120	154	182	202	229	301	352	392	404	440	493	557	572	636	690	722	685	765	843	865	857	956	1028	1082
	Sat. Clay	147	183	217	243	273	362	408	452	482	542	585	630	672	723	782	800	796	882	927	975	995	1102	1159	1218

(continued)

Earth Loads in Trench Conditions (lb/linear ft) $W_c = C_d w B_d B_c$

Depth of Cover	Type of Soil	4" Pipe — Width of Trench (ft.)				6" Pipe — Width of Trench (ft.)				8" Pipe — Width of Trench (ft.)				10" Pipe — Width of Trench (ft.)				12" Pipe — Width of Trench (ft.)				15" Pipe — Width of Trench (ft.)			
		0.75	1.00	1.25	1.5	1.0	1.5	2.0	2.5	1.5	2.0	2.5	3.0	2.0	2.5	3.0	3.5	2.0	2.5	3.0	3.5	2.0	2.5	3.0	3.5
12	Granular w/o Cohesion	68	91	112	132	136	196	246	288	255	320	374	418	400	468	523	565	479	561	627	678	599	701	784	848
	Sand and Soil	87	116	141	164	173	244	303	345	332	405	465	515	505	578	636	676	590	673	754	845	738	842	942	1006
	Sat. Top Soil	102	133	162	185	199	275	325	388	372	456	516	568	566	640	705	755	662	777	841	903	827	953	1051	1129
	Dry Clay	121	158	190	218	235	325	379	432	435	512	575	625	635	718	773	820	739	842	941	985	924	1052	1148	1232
	Sat. Clay	153	194	228	260	289	388	462	501	518	613	666	708	757	825	872	924	902	978	1058	1114	1127	1222	1327	1392
15	Granular w/o Cohesion	68	91	114	134	136	200	257	307	260	335	400	457	418	499	571	625	502	599	685	750	627	749	857	937
	Sand and Soil	87	116	145	170	173	254	322	378	349	436	512	572	542	632	710	766	628	738	828	903	785	922	1035	1129
	Sat. Top Soil	102	136	167	196	202	292	344	410	397	492	572	635	610	710	790	863	704	827	926	998	880	1034	1152	1247
	Dry Clay	121	161	198	231	240	343	421	474	468	571	642	684	700	797	790	943	820	924	1053	1114	1025	1155	1285	1392
	Sat. Clay	155	203	242	275	303	410	503	578	558	683	772	830	820	955	1023	1100	981	1127	1193	1290	1227	1409	1492	1613
18	Granular w/o Cohesion	68	91	114	137	136	204	262	320	265	340	417	479	425	521	599	669	510	625	719	803	638	781	899	1004
	Sand and Soil	87	116	145	174	173	259	334	400	349	436	520	595	548	650	743	818	651	780	892	982	813	975	1115	1227
	Sat. Top Soil	102	136	169	200	202	298	385	494	397	500	577	662	626	721	827	913	751	865	992	1096	938	1081	1240	1370
	Dry Clay	122	162	200	237	242	353	446	518	468	579	673	747	724	842	933	1017	869	1010	1120	1221	1086	1262	1400	1526
	Sat. Clay	156	205	251	291	306	433	530	620	564	690	807	902	862	1008	1127	1199	1034	1210	1353	1439	1293	1513	1691	1798
20	Granular w/o Cohesion	68	91	114	137	136	204	262	327	265	347	425	490	435	531	612	690	520	638	734	828	650	797	918	1035
	Sand and Soil	87	116	145	174	173	259	337	407	349	438	529	606	561	622	757	851	657	794	909	1021	822	992	1136	1276
	Sat. Top Soil	102	136	169	202	202	300	391	459	397	508	596	680	635	745	851	951	762	895	1021	1141	953	1118	1275	1426
	Dry Clay	122	162	200	242	242	358	458	541	468	596	704	783	745	880	979	1057	894	1056	1175	1268	1117	1320	1469	1585
	Sat. Clay	156	205	254	294	309	439	551	646	570	716	840	941	895	1050	1177	1276	1074	1260	1412	1532	1343	1575	1765	1914
25	Granular w/o Cohesion	68	91	114	137	136	204	272	333	265	354	434	510	442	542	638	729	530	650	765	875	663	813	957	1093
	Sand and Soil	87	116	145	174	173	259	345	421	349	449	548	640	561	685	799	900	673	822	959	1080	842	1027	1199	1350
	Sat. Top Soil	102	136	169	203	202	302	401	489	397	521	635	727	651	794	909	1009	781	953	1091	1211	968	1191	1363	1514
	Dry Clay	122	162	200	243	242	362	477	573	471	620	745	857	775	931	1071	1107	930	1117	1285	1328	1163	1396	1607	660
	Sat. Clay	156	208	260	308	309	459	585	688	597	760	895	1021	950	1119	1276	1412	1140	1343	1532	1694	1425	1678	1915	2117
30	Granular w/o Cohesion	68	91	114	137	136	204	272	340	265	354	442	520	442	553	650	744	530	663	780	893	663	829	975	1116
	Sand and Soil	87	116	145	174	173	259	345	431	349	449	561	657	561	701	823	936	673	842	986	1123	842	1052	1233	1404
	Sat. Top Soil	102	136	169	203	202	302	403	496	397	524	645	762	655	807	953	1061	786	968	1144	1273	982	1209	1429	1591
	Dry Clay	122	162	200	243	242	362	477	588	471	620	765	894	775	956	1117	1250	930	1148	1340	1499	1163	1434	1675	1874
	Sat. Clay	156	208	260	311	309	464	605	722	603	787	939	1074	983	1172	1343	1509	1180	1409	1611	1810	1475	1761	2014	2262

Source: Certain-Teed Corporation.

Prism Load (lb/linear ft) $W_c = wHB_c$

Height of Cover Feet	Soil Wt. lb/ft³	Pipe Diameter (Inches)						Height of Cover Feet	Soil Wt. lb/ft³	Pipe Diameter (Inches)					
		4	6	8	10	12	15			4	6	8	10	12	15
3	100	105	157	210	263	313	383	15	100	527	785	1,050	1,313	1,563	1,913
	110	116	173	231	289	343	421		110	580	863	1,155	1,444	1,719	2,105
	120	126	188	252	315	375	459		120	632	941	1,260	1,575	1,876	2,275
	130	137	204	273	341	406	497		130	684	1,020	1,365	1,706	2,032	2,487
4	100	141	209	280	350	417	510	16	100	562	837	1,120	1,400	1,667	2,040
	110	155	230	308	385	458	561		110	618	920	1,232	1,540	1,834	2,245
	120	169	251	336	420	500	612		120	674	1,004	1,344	1,680	2,001	2,448
	130	183	272	364	455	542	663		130	730	1,088	1,456	1,820	2,167	2,653
5	100	176	262	350	438	521	638	17	100	597	889	1,190	1,488	1,771	2,168
	110	193	288	385	481	573	702		110	657	978	1,309	1,636	1,949	2,385
	120	211	314	420	525	625	765		120	716	1,067	1,428	1,785	2,126	2,601
	130	228	340	455	569	677	829		130	776	1,156	1,547	1,934	2,303	2,819
6	100	211	314	420	525	625	765	18	100	632	941	1,260	1,575	1,876	2,295
	110	232	345	462	578	688	842		110	696	1,035	1,386	1,733	2,063	2,525
	120	253	377	504	630	750	918		120	758	1,130	1,512	1,890	2,251	2,754
	130	274	408	546	683	813	995		130	821	1,224	1,638	2,048	2,438	2,984
7	100	246	366	490	613	729	893	19	100	667	994	1,330	1,663	1,980	2,423
	110	271	403	539	674	802	982		110	734	1,093	1,463	1,829	2,178	2,666
	120	295	439	588	735	875	1,071		120	800	1,192	1,596	1,995	2,376	2,907
	130	319	476	637	796	948	1,161		130	867	1,292	1,729	2,161	2,574	3,150
8	100	281	418	560	700	834	1,020	20	100	702	1,046	1,400	1,750	2,084	2,550
	110	309	460	616	770	917	1,122		110	773	1,150	1,540	1,925	2,292	2,806
	120	337	502	672	840	1,000	1,224		120	842	1,255	1,680	2,100	2,501	3,060
	130	365	544	728	910	1,084	1,326		130	913	1,360	1,820	2,275	2,709	3,316
9	100	316	471	630	788	938	1,148	21	100	747	1,098	1,470	1,838	2,188	2,678
	110	348	518	693	866	1,032	1,263		110	812	1,208	1,617	2,021	2,407	2,946
	120	379	565	756	945	1,125	1,377		120	885	1,318	1,764	2,205	2,626	3,213
	130	411	612	819	1,024	1,219	1,492		130	958	1,428	1,911	2,389	2,845	3,482
10	100	351	523	700	875	1,042	1,275	22	100	772	1,151	1,540	1,925	2,292	2,805
	110	387	575	770	963	1,146	1,403		110	850	1,265	1,694	2,117	2,522	3,087
	120	421	628	840	1,050	1,250	1,530		120	927	1,381	1,848	2,310	2,751	3,366
	130	456	680	910	1,138	1,355	1,658		130	1,004	1,496	2,002	2,503	2,980	3,648
11	100	386	575	770	963	1,146	1,403	23	100	807	1,203	1,610	2,013	2,397	2,933
	110	425	633	847	1,059	1,261	1,543		110	889	1,323	1,771	2,214	2,636	3,227
	120	463	690	924	1,155	1,375	1,683		120	969	1,444	1,932	2,415	2,876	3,519
	130	502	748	1,001	1,251	1,490	1,824		130	1,049	1,564	2,093	2,616	3,116	3,813
12	100	421	628	840	1,050	1,251	1,530	24	100	842	1,255	1,680	2,100	2,501	3,060
	110	464	690	924	1,155	1,375	1,684		110	927	1,381	1,848	2,310	2,751	3,367
	120	505	753	1,008	1,260	1,500	1,836		120	1,011	1,506	2,016	2,520	3,001	3,672
	130	548	816	1,092	1,365	1,626	1,990		130	1,095	1,632	2,184	2,730	3,251	3,979
13	100	456	680	910	1,138	1,355	1,658	25	100	878	1,308	1,750	2,187	2,605	3.188
	110	503	748	1,001	1,251	1,490	1,824		110	966	1,438	1,925	2,406	2,866	3,508
	120	548	816	1,092	1,365	1,626	1,989		120	1,053	1,569	2,100	2,625	3,126	3,825
	130	593	884	1,183	1,479	1,761	2,155		130	1,141	1,700	2,275	2,844	3,387	4,145
14	100	491	732	980	1,225	1,459	1,785								
	110	541	805	1,078	1,348	1,605	1,964								
	120	590	879	1,176	1,470	1,751	2,142								
	130	639	952	1,274	1,593	1,896	2,321								

NOTE: Calculations based on actual OD from TABLE 1 ASTM D3034.

F

Appendix

Plastic Pipe Materials Classification

Basic

Abbreviations of Terms Relating to:

D 1600-86 Plastics

Definitions of Terms Relating to:

F 412-87 Plastic Piping Systems

*D 883-86 Plastics

Symbols for:

D2749-68 Dimensions of Plastic Pipe Fittings

Specifications for:

D1788-81 Acrylonitrile–Butadiene–Styrene Plastic (ABS) Plastics, Rigid

D 707-84 Cellulose Acetate Butyrate Molding and Extrusion Compounds

C 582-87 Contact-Molded Reinforced, Thermosetting Plastic (RTP) Laminates for Corrosion Resistant Equipment

D3299-81 Filament-Wound Glass-Fiber-Reinforced Thermoset Resin Chemical-Resistant Tanks

D2581-80 Polybutylene (PB) Plastics Molding and Extrusion Materials

D1248-84 Polyethylene Plastics Molding and Extrusion Materials

D3350-84 Polyethylene Plastics Pipe and Fittings Materials

*D1784-81 Rigid Polyvinyl Chloride (PVC) Compounds and Chlorinated Polyvinyl Chloride (CPVC) Compounds

D2474-81 Vinyl Chloride Copolymer Resins

Acrylonitrile–Butadiene–Styrene (ABS) Plastic Pipe and Fittings

Specifications for:

D2680-87 Acrylonitrile–Butadiene–Styrene (ABS) Composite Sewer Piping

*D2661-87 Acrylonitrile–Butadiene–Styrene (ABS) Plastic Drain, Waste, and Vent Pipe and Fitting

F 628-85 Acrylonitrile–Butadiene–Styrene (ABS) Schedule 40 Plastic Drain, Waste, and Vent Pipe With a Cellular Core

D2468-80 Acrylonitrile–Butadiene–Styrene (ABS) Plastic Pipe Fittings, Schedule 40

D1527-77 Acrylonitrile–Butadiene–Styrene (ABS) Plastic Pipe, Schedules 40 and 80

D2282-82 Acrylonitrile–Butadiene–Styrene (ABS) Plastic Pipe (SDR-PR)

D2750-85 Acrylonitrile–Butadiene–Styrene (ABS) Plastic Utilities Conduit and Fittings

D2751-83 Acrylonitrile–Butadiene–Styrene (ABS) Sewer Pipe and Fittings

---

*Approved for use by agencies of the Department of Defense.

146

D3965-80 Rigid Acrylonitrile–Butadiene–Styrene (ABS) Compounds for Pipe and Fittings

D2235-81 Solvent Cement for Acrylonitrile–Butadiene–Styrene (ABS) Plastic Pipe and Fittings

D3138-83 Solvent Cements for Transition Joints Between Acrylonitrile–Butadiene–Styrene (ABS) and Polyvinyl Chloride (PVC) Non-Pressure Piping Components

F 409-81 Thermoplastic Accessible and Replaceable Plastic Tube and Tubular Fittings

Chlorinated Polyvinyl Chloride (CPVC) Plastic Pipe, Tubing, and Fittings

Specifications for:

*D2846-86 Chlorinated Polyvinyl Chloride (CPVC) Plastic Hot- and Cold-Water Distribution Systems

F 441-86 Chlorinated Polyvinyl Chloride (CPVC) Plastic Pipe, Schedules 40 and 80

F 442-87 Chlorinated Polyvinyl Chloride (CPVC) Plastic Pipe (SDR-PR)

F 438-82 Socket-Type Chlorinated Polyvinyl Chloride (CPVC) Plastic Pipe Fittings, Schedule 40

*F 439-87 Socket-Type Chlorinated Polyvinyl Chloride (CPVC) Plastic Pipe Fittings, Schedule 80

F 437-82 Threaded Chlorinated Polyvinyl Chloride (CPVC) Plastic Pipe Fittings, Schedule 80

Metal Pipe and Fittings, Plastic Lined

Specifications for:

F 781-82 Perfluoro (Alkoxyalkane) Copolymer (PFA) Plastic-Lined Ferrous Metal Pipe and Fittings

F 546-77 Perfluoro (Ethylene-Propylene) Copolymer (FEP) Plastic-Lined Ferrous Metal Pipe and Fittings

F 423-82 Polytetrafluoroethylene (PTFE) Plastic-Lined Ferrous Metal Pipe and Fittings

F 491-77 Polyvinylidene Fluoride (PVDF) Plastic-Lined Ferrous Metal Pipe and Fittings

F 492-85 Propylene and Polypropylene (PP) Plastic-Lined Ferrous Metal Pipe and Fittings

Polybutylene (PB) Plastic Pipe and Tubing

Specifications for:

F 809-83 Large Diameter Polybutylene Plastic Pipe

F 809M-83 Large Diameter Polybutylene Plastic Pipe (Metric)

F 845-86 Plastic Insert Fittings for Polybutylene (PB) Tubing

D2662-83 Polybutylene (PB) Plastic Pipe (SDR-PR)

D3000-73 Polybutylene (PB) Plastic Pipe (SDR-PR) Based on Outside Diameter

D2666-83 Polybutylene (PB) Plastic Tubing

F878-84 Polybutylene (PB) Thermoplastic Thin-Wall Drip Irrigation Tubing

Practice for:

F 699-81 Accelerated Conditioning of Polybutylene Pipe and Tubing for Subsequent Quality Control Testing

Polyethylene Plastic Pipe, Tubing, and Fittings

Specifications for:

D3287-73 Biaxially Oriented Polyethylene (PEO) Plastic Pipe (SDR-PR) Based on Controlled Outside Diameter

D3261-87 Butt Heat Fusion Polyethylene (PE) Plastic Fittings for Polyethylene (PE) Plastic Pipe and Tubing

F 405-85 Corrugated Polyethylene (PE) Tubing and Fittings

F 877-85 Crosslinked Polyethylene (PEX) Plastic Hot- and Cold-Water Distribution Systems

F 876-85 Crosslinked Polyethylene (PEX) Tubing

*D2609-87 Plastic Insert Fittings for Polyethylene (PE) Plastic Pipe

F 892-84 Polyethylene (PE) Corrugated Pipe with a Smooth Interior and Fittings

F 894-85 Polyethylene (PE) Large Diameter Profile Wall Sewer and Drain Pipe

D2104-85	Polyethylene (PE) Plastic Pipe, Schedule 40	F 794-86	Polyvinyl Chloride (PVC) Large Diameter Ribbed Gravity Sewer Pipe and Fittings Based on Controlled Inside Diameter
*D2239-85	Polyethylene (PE) Plastic Pipe (SIDR-PR) Based on Controlled Inside Diameter	D2665-87	Polyvinyl Chloride (PVC) Plastic Drain, Waste, and Vent Pipe and Fittings
D3350-84	Polyethylene Plastics Pipe and Fittings Materials	D2466-83	Polyvinyl Chloride (PVC) Plastic Pipe Fittings, Schedule 40
F 714-85	Polyethylene (PE) Plastic Pipe (SDR-PR) Based on Outside Diameter	D1785-86	Polyvinyl Chloride (PVC) Plastic Pipe, Schedules 40, 80, and 120
D3035-85	Polyethylene (PE) Plastics Pipe (SDR-PR) Based on Controlled Outside Diameter	D2241-87	Polyvinyl Chloride (PVC) Pressure-Rated Pipe (SDR-Series)
D2447-85	Polyethylene (PE) Plastic Pipe, Schedules 40 and 80 Based on Outside Diameter	D2740-83	Polyvinyl Chloride (PVC) Plastic Tubing
*D2737-85	Polyethylene (PE) Plastic Tubing	D2729-85	Polyvinyl Chloride (PVC) Sewer Pipe and Fittings
F 771-85	Polyethylene (PE) Thermoplastic High-Pressure Irrigation Pipeline Systems	F 599-78	Polyvinylidene Chloride (PVDC) Plastic-Lined Ferrous Metal Pipe and Fittings
D2683-87	Socket-Type Polyethylene Fittings for Outside Diameter-Controlled Polyethylene Pipe and Tubing	F 656-80	Primers for Use in Solvent Cement Joints of Polyvinyl Chloride (PVC) Plastic Pipe and Fittings
F 810-85	Smoothwall Polyethylene (PE) Pipe for Use in Drainage and Waste Disposal Absorption Fields	D3678-86	Rigid Polyvinyl Chloride (PVC) Profile Extrusions
		D3679-86	Rigid Polyvinyl Chloride (PVC) Siding

Practice for:

F 905-84	Qualification of Polyethylene Saddle Fusion Joints	D4477-85	Rigid Polyvinyl Chloride (PVC) Soffit
D4101-82	Propylene Plastic Injection and Extrusion Materials	F 512-84	Smooth-Wall Polyvinyl Chloride (PVC) Conduit and Fittings for Underground Installation
		D2467-87	Socket-Type Polyvinyl Chloride (PVC) Plastic Pipe Fittings, Schedule 80

Polyvinyl Chloride (PVC) Plastic Pipe, Tubing, and Fittings

Specifications for:

F 891-86	Coextruded Polyvinyl Chloride (PVC) Plastic Pipe With a Cellular Core	D2672-86	Solvent Cement Joint Sockets on Belled PVC Pressure Pipe
F 800-84	Corrugated Polyvinyl Chloride Tubing and Compatible Fittings	D3138-83	Solvent Cements for Transition Joints Between Acrylonitrile–Butadiene–Styrene (ABS) Polyvinyl Chloride (PVC) Non-Pressure Piping Components
D3915-80	Polyvinyl Chloride (PVC) and Related Plastic Pipe and Fitting Compounds	D2564-84	Solvent Cements for Polyvinyl Chloride (PVC) Plastic Pipe and Fittings
F 949-86	Polyvinyl Chloride (PVC) Corrugated Sewer Pipe With a Smooth Interior and Fittings	F 758-82	Smooth-Wall Polyvinyl Chloride (PVC) Plastic Underdrain Systems for Highway, Airport, and Similar Drainage
F 679-86	Polyvinyl Chloride (PVC) Large-Diameter Plastic Gravity Sewer Pipe and Fittings	F 409-81	Thermoplastic Accessible and

Replaceable Plastic Tube and
Tubular Fittings

D2464-76   Threaded Polyvinyl Chloride
(PVC) Plastic Pipe Fittings,
Schedule 80

D2949-87   3.25-in. Outside Diameter
Polyvinyl Chloride (PVC)
Plastic Drain, Waste, and Vent
Pipe and Fittings

F 789-85   Type PS-46 Polyvinyl Chloride
(PVC) Plastic Gravity Flow
Sewer Pipe and Fittings

D3034-85   Type PSM Polyvinyl Chloride
(PVC) Sewer Pipe and Fittings

D3033-85   Type PSP Polyvinyl Chloride
(PVC) Sewer Pipe and Fittings

Test Methods for:

D2152-80   Degree of Fusion of Extruded
Polyvinyl Chloride (PVC) Pipe
and Molded Fittings by
Acetone Immersion

D4495-85   Impact Resistance of Polyvinyl
Chloride (PVC) Rigid Profiles
by Means of a Falling Weight

Practices for:

F 610-83   Estimating the Quality of
Molded Polyvinyl Chloride
(PVC) Plastic Pipe Fittings by
the Heat Reversion Technique

D2855-83   Making Solvent Cemented Joints
with Polyvinyl Chloride (PVC)
Pipe and Fittings

Resin Pipe, Tubing, and Fittings (Reinforced Thermosetting)

Specifications for:

D4161-86   Bell- and Spigot-Reinforced
Thermosetting Resin Pipe
Joints Using Flexible
Elastomeric Seals

D2997-84   Centrifugally Cast Reinforced
Thermosetting Resin Pipe

D3982-81   Custom Contact-Pressure-Molded
Glass-Fiber-Reinforced
Thermosetting Resin Hoods

D2996-83   Filament Wound Reinforced
Thermosetting Resin Pipe

D2310-80   Machine-Made Reinforced
Thermosetting Resin Pipe,
Classification for

D2517-81   Reinforced Epoxy Resin Gas
Pressure Pipe and Fittings

D3840-81   Reinforced Plastic Mortar Pipe
Fittings for Nonpressure
Applications

D3754-86   Reinforced Plastic Mortar Sewer
and Industrial Pressure Pipe

D4160-82   Reinforced Thermosetting Resin
Pipe (RTRP) Fittings for
Nonpressure Applications

D4163-82   Reinforced Thermosetting Resin
Pressure Pipe (RTRP)

D4024-87   Reinforced Thermosetting Resin
(RTR) Flanges

D4162-82   Reinforced Thermosetting Resin
Sewer and Industrial Pressure
Pipe (RTRP)

D4184-82   Reinforced Thermosetting Resin
Sewer Pipe (RTRP)

D1694-87   Threads for Thermosetting Resin
Pipe

Test Methods for:

D2925-70   Beam Deflection of Reinforced
Thermosetting Plastic Pipe
Under Full Bore Flow,
Measuring

D4398-84   Chemical Resistance of
Fiberglass-Reinforced
Thermosetting Resins by One-
Side Panel Exposure,
Determining

D3615-87   Chemical Resistance of
Thermoset Molding
Compounds Used in the
Manufacture of Molded
Fittings

D2143-69   Cyclic Pressure Strength of
Reinforced Thermosetting
Plastic Pipe

D2105-85   Longitudinal Tensile Properties
of Reinforced Thermosetting
Resin Pipe and Tube

D2992-87   Obtaining Hydrostatic Design
Basis for Reinforced
Thermosetting Resin Pipe and
Fittings

Styrene-Rubber Plastic Pipe and Fittings

Specifications for:

D3122-80   Solvent Cements for Styrene-
Rubber Plastic Pipe and
Fittings

D3298-81   Styrene-Rubber (SR) Plastic
Drain Pipe, Perforated

D2852-81   Styrene-Rubber (SR) Plastic
             Drain Pipe and Fittings

Plastic Pipe Systems Classification

Drain, Waste, and Vent (DWV) Pipe and Fittings

Specifications for:

*D2661-87   Acrylonitrile–Butadiene–Styrene
              (ABS) Plastic Drain, Waste,
              and Vent Pipe and Fittings

F 628-87   Acrylonitrile–Butadiene–Styrene
              (ABS) Schedule 40 Plastic
              Drain, Waste, and Vent Pipe
              With a Cellular Core

D3311-86   Drain, Waste, and Vent (DWV)
              Plastic Fittings Patterns

D2665-87   Polyvinyl Chloride (PVC) Plastic
              Drain, Waste, and Vent Pipe
              and Fittings

F 409-81   Thermoplastic Accessible and
              Replaceable Plastic Tube and
              Tubular Fittings

D2949-87   3.25-in. Outside Diameter
              Polyvinyl Chloride (PVC)
              Plastic Drain, Waste, and Vent
              Pipe and Fittings

Gas Pipe, Tubing, and Fittings

Specifications for:

F 678-82   Polyethylene Gas Pressure Pipe,
              Tubing and Fittings

D2517-81   Reinforced Epoxy Resin Gas
              Pressure Pipe and Fittings

D2513-86   Thermoplastic Gas Pressure
              Piping Systems

Practice for:

F 689-80   Determination of the
              Temperature of Above-Ground
              Plastic Gas Pressure Pipe
              Within Metallic Castings

Sewer Pipe and Fittings

Specifications for:

D2680-87   Acrylonitrile–Butadiene–Styrene
              (ABS) and Polyvinyl Chloride
              (PVC) Composite Sewer Piping

D2751-83   Acrylonitrile–Butadiene–Styrene
              (ABS) Sewer Pipe and Fittings

D3753-81   Glass Fiber-Reinforced Polyester
              Manholes

F 949-86   Polyvinyl Chloride (PVC)
              Corrugated Sewer Pipe With a
              Smooth Interior and Fittings

F 679-86   Polyvinyl Chloride (PVC) Large-
              Diameter Plastic Gravity Sewer
              Pipe and Fittings

D2729-85   Polyvinyl Chloride (PVC) Sewer
              Pipe and Fittings

D3840-81   Reinforced Plastic Mortar Pipe
              Fittings for Nonpressure
              Applications

D3754-86   Reinforced Plastic Mortar Sewer
              and Industrial Pressure Pipe

D3262-87   Reinforced Plastic Mortar Sewer
              Pipe

D4162-82   Reinforced Thermosetting Resin
              Sewer and Industrial Pressure
              Pipe (RTRP)

D4184-82   Reinforced Thermosetting Resin
              Sewer Pipe (RTRP)

D2852-81   Styrene-Rubber (SR) Plastic
              Drain Pipe and Fittings

F 789-85   Type PS-46 Polyvinyl Chloride
              (PVC) Plastic Gravity Flow
              Sewer Pipe and Fittings

D3034-85   Type PSM Polyvinyl Chloride
              (PVC) Sewer Pipe and Fittings

D3033-85   Type PSP Polyvinyl Chloride
              (PVC) Sewer Pipe and Fittings

Practice for:

D2321-83   Underground Installation of
              Flexible Thermoplastic Sewer
              Pipe

Water Pipe

Specifications for:

D2846-86   Chlorinated Polyvinyl Chloride
              (CPVC) Plastic Hot- and Cold-
              Water Distribution Systems

F 877-85   Crosslinked Polyethylene (PEX)
              Plastic Hot- and Cold-Water
              Distribution Systems

D3309-85   Polybutylene (PB) Plastic Hot
              Water Distribution Systems

D3517-83   Reinforced Plastic Mortar
              Pressure Pipe

F 480-81   Thermoplastic Water Well
              Casing and Couplings Made in
              Standard Dimension Ratios
              (SDR)

Practices for:

D2774-72   Underground Installation of
            Thermoplastic Pressure Piping

F 690-86   Underground Installation of
            Thermoplastic Pressure Piping
            Irrigation Systems

Guide for:

F 645-80   Selection, Design, and
            Installation of Thermoplastic
            Water Pressure Piping Systems

Plastic Pipe Installation Components and Procedures

Fittings

Specifications for:

D2468-80   Acrylonitrile–Butadiene–Styrene
            (ABS) Plastic Pipe Fittings,
            Schedule 40

D3261-87   Butt Heat Fusion Polyethylene
            (PE) Plastic Fittings for
            Polyethylene (PE) Plastic Pipe
            and Tubing

D2609–87   Plastic Insert Fittings for
            Polyethylene (PE) Plastic Pipe

D2466-78   Polyvinyl Chloride (PVC) Plastic
            Pipe Fittings, Schedule 40

F 438-82   Socket-Type Chlorinated
            Polyvinyl Chloride (CPVC)
            Plastic Pipe Fittings, Schedule
            40

F 439-87   Socket-Type Chlorinated
            Polyvinyl Chloride (CPVC)
            Plastic Pipe Fittings, Schedule
            80

D2683-87   Socket-Type Polyethylene
            Fittings for Outside Diameter-
            Controlled Polyethylene Pipe
            and Tubing

D2467-87   Socket-Type Polyvinyl Chloride
            (PVC) Plastic Pipe Fittings,
            Schedule 80

F 437-82   Threaded Chlorinated Polyvinyl
            Chloride (CPVC) Plastic Pipe
            Fittings, Schedule 80

D2464-76   Threaded Polyvinyl Chloride
            (PVC) Plastic Pipe Fittings,
            Schedule 80

Practice for:

F 725-81   Drafting Impact Test

Requirements in Thermoplastic
Pipe and Fittings Standards

Joints and Seals

Specifications for:

D4161-86   Bell and Spigot-Reinforced
            Thermosetting Resin Pipe
            Joints Using Flexible
            Elastomeric Seals

F 477-76   Elastomeric Seals (Gaskets) for
            Joining Plastic Pipe

D3212-86   Joints for Drain and Sewer
            Plastic Pipes Using Flexible
            Elastomeric Seals

D3139-84   Joints for Plastic Pressure Pipes
            Using Flexible Elastomeric
            Seals

F 656-80   Primers for Use in Solvent
            Cement Joints of Polyvinyl
            Chloride (PVC) Plastic Pipe
            and Fittings

F 545-80   PVC and ABS Injected Solvent
            Cemented Plastic Pipe Joints

Practices for:

D3140-85   Flaring Polyolefin Pipe and
            Tubing

D2657-87   Heat Joining of Polyolefin Pipe
            and Fittings

D2855-83   Making Solvent-Cemented Joints
            with Polyvinyl Chloride (PVC)
            Pipe and Fittings

Solvent Cement

Specifications for:

D3138-83   Solvent Cements for Transition
            Joints Between Acrylonitrile–
            Butadiene–Styrene (ABS) and
            Polyvinyl Chloride (PVC) Non-
            Pressure Piping Components

D2235-81   Solvent Cements for
            Acrylonitrile–Butadiene–Styrene
            (ABS) Plastic Pipe and Fittings

D2560-80   Solvent Cements for Cellulose
            Acetate Butyrate (CAB) Plastic
            Pipe, Tubing, and Fittings

F 493-85   Solvent Cements for Chlorinated
            Polyvinyl Chloride (CPVC)
            Plastic Pipe and Fittings

D2564-84   Solvent Cements for Polyvinyl
            Chloride (PVC) Plastic Pipe and
            Fittings

D3122-80   Solvent Cements for Styrene-Rubber Plastic Pipe and Fittings

Practice for:

F 402-80   Safe Handling of Solvent Cements and Primers Used for Joining Thermoplastic Pipe and Fittings

Specification for:

F 667-85   8, 10, 12, 15-in. Corrugated Polyethylene Tubing and Fittings

Underground Installation

Practices for:

F 585-78   Insertion of Flexible Polyethylene Pipe into Existing Sewers

F 481-87   Installation of Thermoplastic Pipe and Corrugated Tubing in Septic Tank Leach Fields

F 449-85   Subsurface Installation of Corrugated Thermoplastic Tubing for Agricultural Drainage or Water Table Control

D3839-79   Underground Installation of Flexible Reinforced Thermosetting Resin Pipe and Reinforced Plastic Mortar Pipe

D2321-83   Underground Installation of Flexible Thermoplastic Sewer Pipe

D2774-72   Underground Installation of Thermoplastic Pressure Piping

F 690-86   Underground Installation of Thermoplastic Pressure Piping Irrigation Systems

General Test Methods

See individual sections for specific test methods and practices.

Test Methods for:

D2290-87   Apparent Tensile Strength of Ring or Tubular Plastics and Reinforced Plastics by Split Disk Method

D2925-70   Beam Deflection of Reinforced Thermosetting Resin Pipe

Under Full Bore Flow, Measuring

D3681-83   Chemical Resistance of Reinforced Thermosetting Resin Pipe in a Deflected Condition

D2143-69   Cyclic Pressure Strength of Reinforced, Thermosetting Plastic Pipe

D2152-80   Degree of Fusion of Extruded Polyvinyl Chloride (PVC) Pipe and Molded Fittings by Acetone Immersion

D2122-85   Dimensions of Thermoplastic Pipe and Fittings, Determining

D2412-87   External Loading Properties of Plastic Pipe by Parallel-Plate Loading

D2924-86   External Pressure Resistance of Reinforced Thermosetting Resin Pipe

D2586-68   Hydrostatic Compressive Strength of Glass Reinforced Plastic Cylinders

D2444-84   Impact Resistance of Thermoplastic Pipe and Fittings by Means of a Tup (Falling Weight)

D2105-85   Longitudinal Tensile Properties of Reinforced Thermosetting Resin Pipe and Tube

D4166-82   Measurement of Thickness of Nonmagnetic Materials by Means of a Digital Magnetic Intensity Instrument

D2992-87   Obtaining Hydrostatic Design Basis for Reinforced Thermosetting Resin Pipe and Fittings

D2837-85   Obtaining Hydrostatic Design Basis for Thermoplastic Pipe Materials

D1599-86   Short-Time, Hydraulic Failure Pressure of Plastic Pipe, Tubing and Fittings

D1598-86   Time-to-Failure of Plastic Pipe Under Constant Internal Pressure

F 948-85   Time-to-Failure of Plastic Piping Systems and Components Under Constant Internal Pressure with Flow

Practices for:

C 581-83   Chemical Resistance of
            Thermosetting Resins Used in
            Glass Fiber Reinforced
            Structures Intended for Liquid
            Service, Determining

D 2488-84   Description and Identification of
            Soils (Visual-Manual
            Procedure)

D3567-85   Dimensions of Reinforced
            Thermosetting Resin Pipe
            (RTRP) and Fittings,
            Determining

F 600-78   Nondestructive Ultrasonic
            Evaluation of Socket and Butt
            Joints on Thermoplastic Piping

D2487-85   Soils for Engineering Purposes,
            Classification of

G

Appendix

USEFUL FORMULAS

Water—Unless otherwise specified, data on specific weight and volume of water are usually at 60 deg. F., although sometimes they are at 39.2 deg., at which point water has its greatest density. At 32 deg., water weighs 62.42 lb. per cu. ft.; at 60 deg., 62.37; at 100 deg., 62.00; at 200 deg., 60.13.

Steam—The latent heat of steam is 970 Btu per lb. at atmospheric pressure. That is, after water is raised to 212 deg., 970 more Btu must be added per pound to convert the water to steam at 212 deg.

Btu (British thermal unit)—is the amount of heat required to raise the temperature of one pound of water one degree F. in temperature at about 60 deg.

Temperature and Thermometers—The two principal thermometer scales are the Fahrenheit (used in the U.S.A.) and the Centigrade (used in most non-English speaking countries and in most scientific work). In the former (F) the freezing point is at 32, the boiling point at 212; so that there are 180 divisions or degrees between. With the Centigrade (C) thermometer, freezing is at zero, boiling at 100, with 100 divisions between.

To convert Centigrade degrees to Fahrenheit, multiply the C reading by 9, divide by 5, then add 32.

Example—What is 90 deg. C in Fahrenheit degrees? Multiply 90 by 9 and get 810. Divide by 5 and get 162. Add 32 and get 194 deg F.

To convert Fahrenheit degrees to Centigrade, subtract 32, multiply by 5 and divide by 9.

Example—What is 104 deg. F in Centigrade degrees? Subtract 32 from 104 and obtain 72. Multiply by 5 and get 360. Divide by 9 and the answer is 40 deg. C.

Horsepower—One horsepower is equivalent to 33,000 footpounds per minute. A foot pound is the energy needed to raise 1 pound 1 foot vertically, or $\frac{1}{2}$ pound 2 feet and so on. The power needed to produce 1 horsepower is thus equivalent to raising 33,000 lb. 1 foot, 3,300 lb. 10 feet, 33 lb. 1,000 feet in one minute. To determine the theoretical horsepower required to raise water a given height, multiply the gallons to be raised per minute by 8.33 to find the pounds to be raised per minute. This multiplied by the vertical distance between the source and the point to which the water is to be raised gives the theoretical horsepower.

Tanks—To find the capacity of a cylindrical tank square the diameter in feet, multiply by the length in feet and then by .7854. Result is capacity in cubic feet.

Caution—Due to lack of barrel standardization, the number of gallons in a barrel varies for many different liquids. Consequently, it is safer to use cubic feet as a measure rather than barrels.

Example—What is the capacity of an 8 foot diameter 20 foot long cylindrical tank?

Solution—8 squared is 64. 64 × 20 = 1280 and 1280 multiplied by .7854 = 1005.3 cu ft.

WEIGHTS AND MEASURES

Measures of Length

1 mile = 1760 yards = 5280 feet.
1 yard = 3 feet = 36 inches.   1 foot = 12 inches.
1 mil = 0.001 in.   1 fathom = 2 yards = 6 ft.
1 rod = 5.5 yards = 16.5 feet.
1 hand = 4 inches.   1 span = 9 inches.

1 micro-inch = one millionth inch or 0.000001 inch. (1 micron = one millionth meter = 0.00003937 inch.)

Surveyor's Measure

1 mile = 8 furlongs = 80 chains.
1 furlong = 10 chains = 220 yards.
1 chain = 4 rods = 22 yards = 66 feet = 100 links.
1 link = 7.92 inches.

Nautical Measure

1 league = 3 nautical miles.
1 nautical mile = 6080.2 feet = 1.1516 statute mile. (The knot, which is nautical unit of speed, is equivalent to a speed of 1 nautical mile per hour.)
One degree at the equator = 60 nautical miles = 69.096 statute miles. 360 degrees = 21,600 nautical miles = 24,874.5 statute miles = circumference at equator.

Square Measure

1 square mile = 640 acres = 6400 square chains.
1 acre = 10 square chains = 4840 square yards = 43,560 square feet.
1 square chain = 16 square rods = 484 square yards = 4356 square feet.
1 square rod = 30.25 square yards = 272.25 square feet = 625 square links.
1 square yard = 9 square feet.
1 square foot = 144 square inches.
An acre is equal to a square, the side of which is 208.7 feet.

Cubic Measure

1 cubic yard = 27 cubic feet.
1 cubic foot = 1728 cubic inches.
The following cubic measures are also used for wood and masonry:
1 cord of wood = 4 × 4 × 8 feet = 128 cubic feet.
1 perch of masonry = $16^1/_2$ × $1^1/_2$ × 1 foot = $24^3/_4$ cubic feet.

Shipping Measure

For measuring entire internal capacity of a vessel:
1 register ton = 100 cubic feet.
For measurement of cargo:
Approximately 40 cubic feet of merchandise is considered a shipping ton, unless that bulk would weigh more than 2000 pounds, in which case the freight charge may be based upon weight.
40 cubic feet = 32.143 U.S. bushels = 31.16 Imperial bushels.

Dry Measure

1 bushel (U.S. or Winchester struck bushel) = 1.2445 cubic foot = 2150.42 cubic inches.
1 bushel = 4 pecks = 32 quarts = 64 pints.
1 peck = 8 quarts = 16 pints.
1 quart = 2 pints.
1 heaped bushel = $1^1/_4$ struck bushel.
1 cubic foot = 0.8036 struck bushel.
1 British Imperial bushel = 8 Imperial gallons = 1.2837 cubic foot = 2218.19 cubic inches.

Liquid Measure

1 U.S. gallon = 0.1337 cubic foot = 231 cubic inches = 4 quarts = 8 pints.
1 quart = 2 pints = 8 gills.
1 pint = 4 gills.
1 British Imperial gallon = 1.2009 U.S. gallon = 277.42 cubic inches.
1 cubic foot = 7.48 U.S. gallons.

Apothecaries' Fluid Measure

1 U.S. fluid ounce = 8 drachms = 1.805 cubic inch = 1/128 U.S. gallon.
1 fluid drachm = 60 minims.
1 British fluid ounce = 1.732 cubic inch.

Old Liquid Measure

1 tun = 2 pipes = 3 puncheons.
1 pipe or butt = 2 hogsheads = 4 barrels = 126 gallons.
1 puncheon = 2 tierces = 84 gallons.
1 hogshead = 2 barrels = 63 gallons.
1 tierce = 42 gallons.
1 barrel = $31^1/_2$ gallons.

Avoirdupois or Commercial Weight

1 gross or long ton = 2240 pounds.
1 net or short ton = 2000 pounds.
1 pound = 16 ounces = 7000 grains.
1 ounce = 16 drachms = 437.5 grains.
The following measures for weight are now seldom used in the United States:
1 hundred-weight = 4 quarters = 112 pounds (1 gross or long ton = 20 hundred-weights);

1 quarter = 28 pounds; 1 stone = 14 pounds; 1 quintal = 100 pounds.

Apothecaries' Weight

1 pound = 12 ounces = 5760 grains.
1 ounce = 8 drachms = 480 grains.
1 drachm = 3 scruples = 60 grains.
1 scruple = 20 grains.

Measures of Pressure

1 pound per square inch = 144 pounds per square foot = 0.068 atmosphere = 2.042 inches of mercury at 62 degrees F. = 27.7 inches of water at 62 degrees F. = 2.31 feet of water at 62 degrees F.

1 atmosphere = 30 inches of mercury at 62 degrees F. = 14.7 pounds per square inch = 2116.3 pounds per square foot = 33.95 feet of water at 62 degrees F.

1 foot of water at 62 degrees F. = 62.355 pounds per square foot = 0.433 pound per square inch.

1 inch of mercury at 62 degrees F. = 1.132 foot of water = 13.58 inches of water = 0.491 pound per square inch.

Conversion Factors

Multiply	By	To Obtain
Atmosphere	29.92	Inches of mercury
Atmosphere	33.90	Feet of water
Atmosphere	14.70	Pounds per square in.
Barrels (oil)	42	Gallons
Boiler horsepower	33,475	Btu per hour
Btu	0.252	Calories
Centimeters	0.3937	Inches
Cubic feet	1728	Cubic inches
Cubic feet of water	7.48	Gallons of water
Cubic feet of water	62.37	Pounds of water
Cubic feet per minute	0.1247	Gallons per second

Conversion Factors (*continued*)

Multiply	By	To Obtain
Cubic meters	35.314	Cubic feet
Cubic meters	264.2	U.S. gallons
Feet	0.3048	Meters
Feet of water	0.881	Inches of mercury
Feet of water	62.37	Pounds per square ft.
Feet of water	0.4335	Pounds per square in.
Feet of water	0.0295	Atmospheres
Feet per minute	0.01136	Miles per hour
Feet per minute	0.01667	Feet per second
Gallons (U.S.)	0.1337	Cubic feet
Gallons (U.S.)	231	Cubic inches
Gallons (U.S.)	8.3453	Pounds of water
Gallons (Imperial)	277.3	Cubic inches
Gallons (Imperial)	10	Pounds of water
Grams	0.03527	Ounces
Horsepower	33,000	Foot-pounds per min.
Horsepower	0.7457	Kilowatts
Horsepower (boiler)	33,471.9	Btu per hour
Inches	2.54	Centimeters
Inches	25.4	Millimeters
Inches of mercury	1.131	Feet of water
Inches of mercury	0.4912	Pounds per square in.
Inches of water	0.03613	Pounds per square in.
Inches of water	5.202	Pounds per square ft.
Kilograms	2.2046	Pounds
Kilometers	0.6214	Miles
Kilowatts	1.341	Horsepower
Kilowatt-hours	3415	Btu
Liters	0.2642	U.S. gallons
Meters	39.37	Inches
Meters	3.2808	Feet
Miles	1.609	Kilometers
Millimeters	0.03937	Inches
Ounces	28.35	Grams
Pounds	0.4536	Kilograms
Pounds	7000	Grains
Pounds of water	0.01602	Cubic feet
Pounds of water	27.68	Cubic inches
Pounds of water	0.1198	Gallons
Pounds per square in.	2.309	Feet of water
Pounds per square in.	2.0416	Inches of mercury
Tons of refrigeration	12,000	Btu per hour
Tons (long)	2240	Pounds
Tons, metric	2204.6	Pounds
Tons (short)	2000	Pounds
Watts	0.5692	Btu per minute
Watt-hours	3.415	Btu

Conversion of Pounds per Square Inch to Head in Feet of Water

Pounds per Sq. In.	Head in Feet	Pounds per Sq. In.	Head in Feet
1	2.31	120	277.07
2	4.62	125	288.62
3	6.93	130	300.16
4	9.24	140	323.25
5	11.54	150	346.34
6	13.85	160	369.43
7	16.16	170	392.52
8	18.47	180	415.61
9	20.78	190	438.90
10	23.09	200	461.78
15	34.63	225	519.51
20	46.18	250	577.24
25	57.72	275	643.03
30	69.27	300	692.69
40	92.36	325	750.41
50	115.45	350	808.13
60	138.54	375	865.89
70	161.63	400	922.58
80	184.72	500	1154.48
90	207.81	1000	2309.00
100	230.90	1500	3463.48
110	253.98	2000	4618.00

Conversion of Head in Feet of Water to Pounds per Square Inch

Head in Feet	Pounds per Sq. In.	Head in Feet	Pounds per Sq. In.
1	.43	140	60.63
2	.87	150	64.96
3	1.30	160	69.29
4	1.73	170	73.63
5	2.17	180	77.96
6	2.60	190	82.29
7	3.03	200	86.62
8	3.46	225	97.45
9	3.90	250	108.27
10	4.33	275	119.10
20	8.66	300	129.93
30	12.99	325	140.75
40	17.32	350	151.58
50	21.65	400	173.24
60	25.99	500	216.55
70	30.32	600	259.85
80	34.65	700	303.16
90	38.98	800	346.47
100	43.31	900	389.78
110	47.65	1000	433.09
120	51.97	1500	649.64
130	56.30	2000	866.18

Fahrenheit-Centigrade Conversion

In the temperature conversion table below, numbers in the center column, in bold face type, refer to the temperature in either Fahrenheit or Centigrade degrees. If it is desired to convert from Fahrenheit to Centigrade degrees, consider the center column as a table of Fahrenheit temperatures and read the corresponding Centigrade temperature in the column at the left. If it is desired to convert from Centigrade to Fahrenheit degrees, consider the center column as a table of Centigrade values, and read the corresponding Fahrenheit temperature on the right.

For example, if it is desired to convert 40 degrees Fahrenheit to Centigrade degrees, the column at left shows 4.4 degrees Centigrade for a reading of 40 (degrees Fahrenheit) in the center column. Interpolation factors are given for use with that portion of the table in which the center column advances in increments of 10. To illustrate, suppose it is desired to find the Fahrenheit equivalent of 314 deg C. The equivalent of 310 deg C, found in the body of the main table, is seen to be 590 deg F. The Fahrenheit equivalent of a 4-deg C difference is seen to be 7.2, as read in the table of interpolating factors. The answer is the sum of the two, or 597.2 deg F.

Deg C		Deg F		Deg C		Deg F
0.56	1	1.8		3.33	6	10.8
1.11	2	3.6		3.89	7	12.6
1.67	3	5.4		4.44	8	14.4
2.22	4	7.2		5.00	9	16.2
2.78	5	9.0		5.56	10	18.0

For conversions not covered in the table, the following formulas are used:

$$F = 1.8\,C + 32$$
$$C = (F - 32) \div 1.8$$

Deg C		Deg F	Deg C		Deg F	Deg C		Deg F	Deg C		Deg F
-17.8	0	32—	0.6	33	91.4	18.9	66	150.8	37.2	99	210.2
-17.2	1	33.8	1.1	34	93.2	19.4	67	152.6	37.8	100	212.0
-16.7	2	35.6	1.7	35	95.0	20.0	68	154.4	38.3	101	213.8
-16.1	3	37.4	2.2	36	96.8	20.6	69	156.2	38.9	102	215.6
-15.6	4	39.2	2.7	37	98.6	21.1	70	158.0	39.4	103	217.4
-15.0	5	41.0	3.3	38	100.4	21.7	71	159.8	40.0	104	219.2
-14.4	6	42.8	3.9	39	102.2	22.2	72	161.6	40.6	105	221.0
-13.9	7	44.6	4.4	40	104.0	22.8	73	163.4	41.1	106	222.8
-13.3	8	46.4	5.0	41	105.8	23.3	74	165.2	41.7	107	224.6
-12.8	9	48.2	5.6	42	107.6	23.9	75	167.0	42.2	108	226.4
-12.2	10	50.0	6.1	43	109.4	24.4	76	168.8	42.8	109	228.2
-11.7	11	51.8	6.7	44	111.2	25.0	77	170.6	43.3	110	230.0
-11.1	12	53.6	7.2	45	113.0	25.6	78	172.4	43.9	111	231.8
-10.6	13	55.4	7.8	46	114.8	26.1	79	174.2	44.4	112	233.6
-10.0	14	57.2	8.3	47	116.6	26.7	80	176.0	45.0	113	235.4
-9.4	15	59.0	8.9	48	118.4	27.2	81	177.8	45.6	114	237.2
-8.9	16	60.8	9.4	49	120.2	27.8	82	179.6	46.1	115	239.0
-8.3	17	62.6	10.0	50	122.0	28.3	83	181.4	46.7	116	240.8
-7.8	18	64.4	10.6	51	123.8	28.9	84	183.2	47.2	117	242.6
-7.2	19	66.2	11.1	52	125.6	29.4	85	185.0	47.8	118	244.4
-6.7	20	68.0	11.7	53	127.4	30.0	86	186.8	48.3	119	246.2
-6.1	21	69.8	12.2	54	129.2	30.6	87	188.6	48.9	120	248.0
-5.6	22	71.6	12.8	55	131.0	31.1	88	190.4	49.4	121	249.8
-5.0	23	73.4	13.3	56	132.8	31.7	89	192.2	50.0	122	251.6
-4.4	24	75.2	13.9	57	134.6	32.2	90	194.0	50.6	123	253.4
-3.9	25	77.0	14.4	58	136.4	32.8	91	195.8	51.1	124	255.2
-3.3	26	78.8	15.0	59	138.2	33.3	92	197.6	51.7	125	257.0
-2.8	27	80.6	15.6	60	140.0	33.9	93	199.4	52.2	126	258.8
-2.2	28	82.4	16.1	61	141.8	34.4	94	201.2	52.8	127	260.6
-1.7	29	84.2	16.7	62	143.6	35.0	95	203.0	53.3	128	262.4
-1.1	30	86.0	17.2	63	145.4	35.6	96	204.8	53.9	129	264.2
-0.6	31	87.8	17.8	64	147.2	36.1	97	206.6	54.4	130	266.0
-0	32	89.6	18.3	65	149.0	36.7	98	208.4	55.0	131	267.8

Deg C		Deg F	Deg C		Deg F	Deg C		Deg F	Deg C		Deg F
55.6	132	269.6	73.9	165	329.0	92.2	198	388.4	204.4	400	752.0
56.1	133	271.4	74.4	166	330.8	92.8	199	390.2	210	410	770.0
56.7	134	273.2	75.0	167	332.6	93.3	200	392.0	215.6	420	788
57.2	135	275.0	75.6	168	334.4	93.9	201	393.8	221.1	430	806
57.8	136	276.8	76.1	169	336.2	94.4	202	395.6	226.7	440	824
58.3	137	278.6	76.7	170	338.0	95.0	203	397.4	232.2	450	842
58.9	138	280.4	77.2	171	339.8	95.6	204	399.2	237.8	460	860
59.4	139	282.2	77.8	172	341.6	96.1	205	401.0	243.3	470	878
60.0	140	284.0	78.3	173	343.4	96.7	206	402.8	248.9	480	896
60.6	141	285.8	78.9	174	345.2	97.2	207	404.6	254.4	490	914
61.1	142	287.6	79.4	175	347.0	97.8	208	406.4	260.0	500	932
61.7	143	289.4	80.0	176	348.8	98.3	209	408.2	265.6	510	950
62.2	144	291.2	80.6	177	350.6	98.9	210	410.0	271.1	520	968
62.8	145	293.0	81.1	178	352.4	99.4	211	411.8	276.7	530	986
63.3	146	294.8	81.7	179	354.2	100.0	212	413.6	282.2	540	1004
63.9	147	296.6	82.2	180	356.0	104.4	220	428.0	287.8	550	1022
64.4	148	298.4	82.8	181	357.8	110.0	230	446.0	293.3	560	1040
65.0	149	300.2	83.3	182	359.6	115.6	240	464.0	298.9	570	1058
65.6	150	302.0	83.9	183	361.4	121.1	250	482.0	304.4	580	1076
66.1	151	303.8	84.4	184	363.2	126.7	260	500.0	310.0	590	1094
66.7	152	305.6	85.0	185	365.0	132.2	270	518.0	315.6	600	1112
67.2	153	307.4	85.6	186	366.8	137.8	280	536.0	321.1	610	1130
67.8	154	309.2	86.1	187	368.6	143.3	290	554.0	326.7	620	1148
68.3	155	311.0	86.7	188	370.4	148.9	300	572.0	332.2	630	1166
68.9	156	312.8	87.2	189	372.2	154.4	310	590.0	337.8	640	1184
69.4	157	314.6	87.8	190	374.0	160.0	320	608.0	343.3	650	1202
70.0	158	316.4	88.3	191	375.8	165.6	330	626.0	348.9	660	1220
70.6	159	318.2	88.9	192	377.6	171.1	340	644.0	354.4	670	1238
71.1	160	320.0	89.4	193	379.4	176.7	350	662.0	360.0	680	1256
71.7	161	321.8	90.0	194	381.2	182.2	360	680.0	365.6	690	1274
72.2	162	323.6	90.6	195	383.0	187.8	370	698.0	371.1	700	1292
72.8	163	325.4	91.1	196	384.8	193.3	380	716.0	376.7	710	1310
73.3	164	327.2	91.7	197	386.6	198.9	390	734.0	382.2	720	1328

Specific Gravity of Liquids @ 68° F

Liquid	Specific Gravity
Acetic Acid, 50%	1.06
Acetic Anhydride	1.08
Acetone	.79
Alcohol Amyl	.82
Alcohol, Butyl	.81
Aniline	1.02
Benzene	.88
Brine (25% NaCl)	1.18
Butyl Acetate	.88
Calcium Chloride, 25%	1.23
Carbon Tetrachloride	1.60
Chlorobenzene	1.12
Chromic Acid, 10%	1.07
Chromic Acid, 50%	1.50
Cyclohexanone	.94
Ethylene Bromide	2.18
Ethylene Chloride	1.25
Ferric Chloride, 20%	1.18
Ferric Chloride, 46%	1.50
Fuel Oil No. 1 and 2	.95
Heptane	.68
Hydrogen Peroxide, 30%	1.11
Hydrochloric Acid, 37%	1.18
Kerosene (85° F)	.82
Methyl Ethyl Ketone (MEK)	.81
Nitric Acid, 30%	1.18
Oil, Lubricating SAE 10-20-30 (@ 115° F)	.94
Oleic Acid	.89
Phenol	1.07
Phosphoric Acid, 50%	1.34
Propionic Acid	.99
Pyridine	.98
Sodium Hydroxide, 50%	1.53
Sulfuric Acid, 20%	1.14
Sulfuric Acid, 50%	1.40

Specific Gravity of Liquids @ 68° F (continued)

Liquid	Specific Gravity
Sulfuric Acid, 85%	1.79
Trichloroethylene	1.47
Toluene	.87
Urea	1.36
Water	1.00
Water (Sea)	1.02–1.03
Xylene	.86
Zinc Chloride, 50%	1.61

Courtesy of Chemtrol Div. of Nibco.

Specific Gravity of Gases

(At 60° F and 29.92″ Hg)		
Dry Air (1 cu. ft. at 60° F)		1.000
Acetylene	C_2H_2	0.91
Ethane	C_2H_6	1.05
Methane	CH_4	0.554
Ammonia	NH_3	0.596
Carbon-dioxide	CO_2	1.53
Carbon-monoxide	CO	0.967
Butane	C_4H_{10}	2.067
Butene	C_4H_8	1.93
Chlorine	Cl_2	2.486
Helium	He	0.138
Hydrogen	H_2	0.0696
Nitrogen	N_2	0.9718
Oxygen	O_2	1.1053

Contents of Cylindrical Tanks in U.S. Gallons

Length of Tank, Feet	Diameter of Tank, Feet														
	5	6	7	8	9	10	11	12	14	16	18	20	22	24	25
	Contents of Tank, U.S. Gallons					Contents of Tank, U.S. Gallons					Contents of Tank, U.S. Gallons				
5	734	1058	1439	1880	2379	2938	3555	4230	5758	7521	9518	11751	14218	16921	18360
6	881	1269	1727	2256	2855	3525	4265	5076	6909	9025	11422	14101	17062	20305	22032
7	1028	1481	2015	2632	3331	4113	4976	5922	8061	10529	13325	16451	19905	23689	25704
8	1175	1692	2303	3008	3807	4700	5687	6768	9212	12033	15229	18801	22749	27073	29376
9	1322	1904	2591	3384	4283	5288	6398	7614	10364	13537	17132	21151	25592	30457	33048
10	1469	2115	2879	3760	4759	5875	7109	8460	11515	15041	19036	23501	28436	33841	36720
11	1616	2327	3167	4136	5235	6463	7820	9306	12667	16545	20940	25851	31280	37225	40392
12	1763	2538	3455	4512	5711	7050	8531	10152	13818	18049	22843	28201	34123	40609	44064
13	1909	2750	3742	4888	6187	7638	9242	10998	14970	19553	24747	30551	36967	43993	47736
14	2056	2961	4030	5264	6662	8225	9953	11844	16121	21057	26650	32901	39810	47377	51408
15	2203	3173	4318	5640	7138	8813	10664	12690	17273	22562	28554	35252	42654	50762	55080
16	2350	3384	4606	6016	7614	9400	11374	13536	18424	24066	30458	37602	45498	54146	58752
17	2497	3596	4894	6392	8090	9988	12085	14383	19576	25570	32361	39952	48341	57530	62424
18	2644	3807	5182	6768	8566	10575	12796	15229	20727	27074	34265	42302	51185	60914	66096
19	2791	4019	5480	7144	9042	11163	13507	16075	21879	28578	36168	44652	54028	64298	69768
20	2938	4230	5758	7520	9518	11750	14218	16921	23030	30082	38072	47002	56872	67682	73440

Diameter	Circumference	Area	Diameter	Circumferences	Area	Diameter	Circumference	Area
$1/8$	0.3927	0.0123	8	25.1327	50.265	16	50.2655	201.06
$1/4$	0.7854	0.0491	$1/8$	25.5254	51.849	$1/8$	50.6582	204.22
$3/8$	1.1781	0.1105	$1/4$	25.9181	53.456	$1/4$	51.0509	207.39
$1/2$	1.5708	0.1964	$3/8$	26.3108	55.088	$3/8$	51.4436	210.60
$5/8$	1.9635	0.3068	$1/2$	26.7035	56.745	$1/2$	51.8363	213.82
$3/4$	2.3562	0.4418	$5/8$	27.0962	58.426	$5/8$	52.2290	217.08
$7/8$	2.7489	0.6013	$3/4$	27.4889	60.132	$3/4$	52.6217	220.35
			$7/8$	27.8816	61.862	$7/8$	53.0144	223.65
1	3.1416	0.7854	9	28.2743	63.617	17	53.4071	226.98
$1/8$	3.5343	0.9940	$1/8$	28.6670	65.397	$1/8$	53.7998	230.33
$1/4$	3.9270	1.2272	$1/4$	29.0597	67.201	$1/4$	54.1925	233.71
$3/8$	4.3197	1.4849	$3/8$	29.4524	69.029	$3/8$	54.5852	237.10
$1/2$	4.7124	1.7671	$1/2$	29.8451	70.882	$1/2$	54.9779	240.53
$5/8$	5.1051	2.0739	$5/8$	30.2378	72.760	$5/8$	55.3706	243.98
$3/4$	5.4978	2.4053	$3/4$	30.6305	74.662	$3/4$	55.7633	247.45
$7/8$	5.8905	2.7612	$7/8$	31.0232	76.589	$7/8$	56.1560	250.95
2	6.2832	3.1416	10	31.4159	78.540	18	56.5487	254.47
$1/8$	6.6759	3.5466	$1/8$	31.8086	80.516	$1/8$	56.9414	258.02
$1/4$	7.0686	3.9761	$1/4$	32.2013	82.516	$1/4$	57.3341	261.59
$3/8$	7.4613	4.4301	$3/8$	32.5940	84.541	$3/8$	57.7268	265.18
$1/2$	7.8540	4.9087	$1/2$	32.9867	86.590	$1/2$	58.1195	268.80
$5/8$	8.2467	5.4119	$4/8$	33.3794	88.664	$5/8$	58.5122	272.45
$3/4$	8.6394	5.9396	$3/4$	33.7721	90.763	$3/4$	58.9049	276.12
$7/8$	9.0321	6.4918	$7/8$	34.1648	92.886	$7/8$	59.2976	279.81
3	9.4248	7.0686	11	34.5575	95.03	19	59.6903	283.53
$1/8$	9.8175	7.6699	$1/8$	34.9502	97.20	$1/8$	60.0830	287.27
$1/4$	10.2102	8.2958	$1/4$	35.3429	99.40	$1/4$	60.4757	291.04
$3/8$	10.6029	8.9462	$3/8$	35.7356	101.62	$3/8$	60.8684	294.83
$1/2$	10.9956	9.6211	$1/2$	36.1283	103.87	$1/2$	61.2611	298.65
$5/8$	11.3883	10.3211	$5/8$	36.5210	106.14	$5/8$	61.6538	302.49
$3/4$	11.7810	11.045	$3/4$	36.9137	108.43	$3/4$	62.0465	306.35
$7/8$	12.1737	11.793	$7/8$	37.3064	110.75	$7/8$	62.4392	310.24
4	12.5664	12.566	12	37.6991	113.10	20	62.8319	314.16
$1/8$	12.9591	13.364	$1/8$	38.0918	115.47	$1/8$	63.2246	318.10
$1/4$	13.3518	14.186	$1/4$	38.4845	117.86	$1/4$	63.6173	322.06
$3/8$	13.7445	15.033	$3/8$	38.8772	120.28	$3/8$	64.0100	326.05
$1/2$	14.1372	15.904	$1/2$	39.2699	122.72	$1/2$	64.4026	330.06
$5/8$	14.5299	16.800	$5/8$	39.6626	125.19	$5/8$	64.7953	334.10
$3/4$	14.9226	17.721	$3/4$	40.0553	127.68	$3/4$	65.1880	338.16
$7/8$	15.3153	18.665	$7/8$	40.4480	130.19	$7/8$	65.5807	342.25
5	15.7080	19.635	13	40.8407	132.73	21	65.9734	346.36
$1/8$	16.1007	20.629	$1/8$	41.2334	135.30	$1/8$	66.3661	350.50
$1/4$	16.4934	21.648	$1/4$	41.6261	137.89	$1/4$	66.7588	354.66
$3/8$	16.8861	22.691	$3/8$	42.0188	140.50	$3/8$	67.1515	358.84
$1/2$	17.2788	23.758	$1/2$	42.4115	143.14	$1/2$	67.5442	363.05
$5/8$	17.6715	24.850	$5/8$	42.8042	145.80	$5/8$	67.9369	367.28
$3/4$	18.0642	25.967	$3/4$	43.1969	148.49	$3/4$	68.3296	371.54
$7/8$	18.4569	27.109	$7/8$	43.5896	151.20	$7/8$	68.7223	375.83
6	18.8496	28.274	14	43.9823	153.94	22	69.1150	380.13
$1/8$	19.2423	29.465	$1/8$	44.3750	156.70	$1/8$	69.5077	384.46
$1/4$	19.6350	30.680	$1/4$	44.7677	159.48	$1/4$	69.9004	388.82
$3/8$	20.0277	31.919	$3/8$	45.1604	162.30	$3/8$	70.2931	393.20
$1/2$	20.4204	33.183	$1/2$	45.5531	165.13	$1/2$	70.6858	397.61
$5/8$	20.8131	34.472	$5/8$	45.9458	167.99	$5/8$	71.0785	402.04
$3/4$	21.2058	35.785	$3/4$	46.3385	170.87	$3/4$	71.4712	406.49
$7/8$	21.5984	37.122	$7/8$	46.7312	173.78	$7/8$	71.8639	410.97
7	21.9911	38.485	15	47.1239	176.71	23	72.2566	415.48
$1/8$	22.3838	39.871	$1/8$	47.5166	179.67	$1/8$	72.6493	420.00
$1/4$	22.7765	41.282	$1/4$	47.9093	182.65	$1/4$	73.0420	424.56
$3/8$	23.1692	42.718	$3/8$	48.3020	185.66	$3/8$	73.4347	429.13
$1/2$	23.5619	44.179	$1/2$	48.6947	188.69	$1/2$	73.8274	433.74
$5/8$	23.9546	45.664	$5/8$	49.0874	191.75	$5/8$	74.2201	438.36
$3/4$	24.3473	47.173	$3/4$	49.4801	194.83	$3/4$	74.6128	443.01
$7/8$	24.7400	48.707	$7/8$	49.8728	197.93	$7/8$	75.0055	447.69

Areas of Circles

Diameter	Circumference	Area	Diameter	Circumference	Area	Diameter	Circumference	Area
24	75.3982	452.39	32	100.531	804.25	40	125.664	1256.6
1/8	75.7909	457.11	1/8	100.924	810.54	1/8	126.056	1264.5
1/4	76.1836	461.86	1/4	101.316	816.86	1/4	126.449	1272.4
3/8	76.5763	466.64	3/8	101.709	823.21	3/8	126.842	1280.3
1/2	76.9690	471.44	1/2	102.102	829.58	1/2	127.235	1288.2
5/8	77.3617	476.26	5/8	102.494	835.97	5/8	127.627	1296.2
3/4	77.7544	481.11	3/4	102.887	842.39	3/4	128.020	1304.2
7/8	78.1471	485.98	7/8	103.280	848.33	7/8	128.413	1312.2
25	78.5398	490.87	33	103.673	855.30	41	128.805	1320.2
1/8	78.9325	495.79	1/8	104.065	861.79	1/8	129.198	1328.3
1/4	79.3252	500.74	1/4	104.458	868.31	1/4	129.591	1336.4
3/8	79.7179	505.71	3/8	104.851	874.85	3/8	129.983	1344.5
1/2	80.1106	510.71	1/2	105.243	881.41	1/2	130.376	1352.7
5/8	80.5033	515.72	5/8	105.636	888.00	5/8	130.769	1360.8
3/4	80.8960	520.77	3/4	106.029	894.62	3/4	131.161	1369.0
7/8	81.2887	525.84	7/8	106.421	901.26	7/8	131.554	1377.2
26	81.6814	530.93	34	106.814	907.92	42	131.947	1385.4
1/8	82.0741	536.05	1/8	107.207	914.61	1/8	132.340	1393.7
1/4	82.4668	541.19	1/4	107.600	921.32	1/4	132.723	1402.0
3/8	82.8595	546.35	3/8	107.992	928.06	3/8	133.125	1410.3
1/2	83.2522	551.55	1/2	108.385	934.82	1/2	133.518	1418.6
5/8	83.6449	556.76	5/8	108.778	941.61	5/8	133.910	1427.0
3/4	84.0376	562.00	3/4	109.170	948.42	3/4	134.303	1435.4
7/8	84.4303	567.27	7/8	109.563	955.25	7/8	134.696	1443.8
27	84.8230	572.56	35	109.956	962.1	43	135.088	1452.2
1/8	85.2157	577.87	1/8	110.348	969.0	1/8	135.481	1460.7
1/4	85.6084	583.21	1/4	110.741	975.9	1/4	135.874	1469.1
3/8	86.0011	588.57	3/8	111.134	982.8	3/8	136.267	1477.6
1/2	86.3938	593.96	1/2	111.527	989.8	1/2	136.659	1486.2
5/8	86.7865	599.37	5/8	111.919	996.8	5/8	137.052	1494.7
3/4	87.1792	604.81	3/4	112.312	1003.8	3/4	137.445	1503.3
7/8	87.5719	610.27	7/8	112.705	1010.8	7/8	137.837	1511.9
28	87.9646	615.75	36	113.097	1017.9	44	138.230	1520.5
1/8	88.3573	621.26	1/8	113.490	1025.0	1/8	136.623	1529.2
1/4	88.7500	626.80	1/4	113.883	1032.1	1/4	139.015	1537.9
3/8	89.1427	632.36	3/8	114.275	1039.2	3/8	139.408	1546.6
1/2	89.5354	637.94	1/2	114.668	1046.3	1/2	139.801	1555.3
5/8	89.9281	643.55	5/8	115.061	1053.5	5/8	140.194	1464.0
3/4	90.3208	649.18	3/4	115.454	1060.7	3/4	140.586	1572.8
7/8	90.7135	654.84	7/8	115.846	1068.0	7/8	140.979	1581.6
29	91.1062	660.52	37	116.239	1075.2	45	141.372	1590.4
1/8	91.4989	666.23	1/8	116.632	1082.5	1/8	141.764	1599.3
1/4	91.8916	672.96	1/4	117.024	1089.8	1/4	142.157	1608.2
3/8	92.2843	677.71	3/8	117.417	1097.1	3/8	142.550	1617.0
1/2	92.6770	683.49	1/2	117.810	1104.5	1/2	142.942	1626.0
5/8	93.0697	689.30	5/8	118.202	1111.8	5/8	143.335	1634.9
3/4	93.4624	695.13	3/4	118.596	1119.2	3/4	143.728	1643.9
7/8	93.8551	700.98	7/8	118.988	1126.7	7/8	144.121	1652.9
30	94.2478	706.86	38	119.381	1134.1	46	144.513	1661.9
1/8	94.6405	712.76	1/8	119.773	1141.6	1/8	144.906	1670.9
1/4	95.0332	718.69	1/4	120.166	1149.1	1/4	145.299	1680.0
3/8	95.4259	724.64	3/8	120.559	1156.6	3/8	145.691	1689.1
1/2	95.8186	730.62	1/2	120.951	1164.2	1/2	146.084	1698.2
5/8	96.2113	736.62	5/8	121.344	1171.7	5/8	146.477	1707.4
3/4	96.6040	742.64	3/4	121.737	1179.2	3/4	146.869	1716.5
7/8	96.9967	748.69	7/8	122.129	1186.9	7/8	147.262	1725.7
31	97.3894	754.77	39	122.522	1194.6			
1/8	97.7821	760.87	1/8	122.915	1202.3			
1/4	98.1748	766.99	1/4	123.308	1210.0			
3/8	98.5675	773.14	3/8	123.700	1217.7			
1/2	98.9602	779.31	1/2	124.093	1225.4			
5/8	99.3529	785.51	5/8	124.486	1233.1			
3/4	99.7456	791.73	3/4	124.878	1241.0			
7/8	100.138	797.98	7/8	125.271	1248.8			

Areas and Volumes

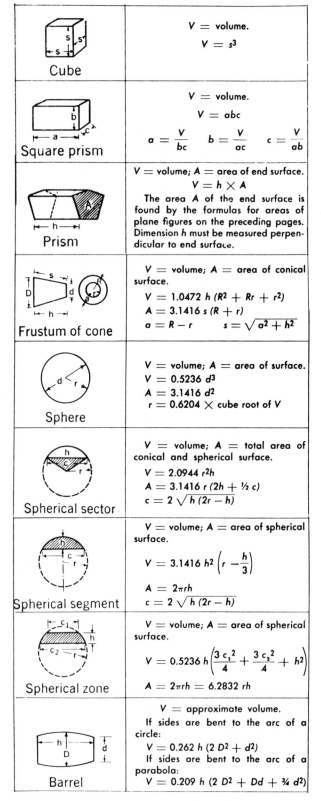

Pyramid

V = volume.

$V = \frac{1}{3} h \times$ area of base.

Frustum of pyramid

V = volume.

$$V = \frac{h}{3}\left(A_1 + A_2 + \sqrt{A_1 \times A_2}\right)$$

Cylinder

V = volume; S = area of curved surface.

$V = 0.7854\, d^2 h$

$S = 6.2832\, rh$

Area of all surfaces is S plus twice the area of an end.

Portion of cylinder

V = volume; S = area of curved surface.

$V = 0.3927\, d^2 (h_1 + h_2)$

$S = 1.5708\, d\,(h_1 + h_2)$

Hollow cylinder

V = volume.

$V = 0.7854\, h\,(D^2 - d^2)$

$\quad = 3.1416\, ht\,(R + r)$

Cone

V = volume; A = area of conical surface.

$V = 0.2618\, d^2 h$

$A = 1.5708\, ds$

$s = \sqrt{r^2 + h^2}$

Circular sector

A = area; l = length of arc;
x = angle, in degrees.

$l = \dfrac{2A}{r}$

$A = \frac{1}{2}\, rl$

$x = \dfrac{57.296\, l}{r} \qquad r = \dfrac{2A}{l}$

Circular segment

A = area; l = length of arc;
x = angle, in degrees.

$c = 2\sqrt{h\,(2r - h)}$

$A = \frac{1}{2}\,[rl - c\,(r - h)]$

$l = 0.01745\, rx$

$h = r - \frac{1}{2}\sqrt{4\, r^2 - c^2}$

$x = \dfrac{57.296\, l}{r}$

Ellipse

A = area;
P = perimeter or circumference.
$A = 3.1416\, ab$.

An approximate formula for the perimeter is:

$P = 3.1416\sqrt{2\,(a^2 + b^2)}$

Areas and Volumes

Cube

V = volume.

$V = s^3$

Square prism

V = volume.

$V = abc$

$a = \dfrac{V}{bc} \qquad b = \dfrac{V}{ac} \qquad c = \dfrac{V}{ab}$

Prism

V = volume; A = area of end surface.

$V = h \times A$

The area A of the end surface is found by the formulas for areas of plane figures on the preceding pages. Dimension h must be measured perpendicular to end surface.

Frustum of cone

V = volume; A = area of conical surface.

$V = 1.0472\, h\,(R^2 + Rr + r^2)$

$A = 3.1416\, s\,(R + r)$

$a = R - r \qquad s = \sqrt{a^2 + h^2}$

Sphere

V = volume; A = area of surface.

$V = 0.5236\, d^3$

$A = 3.1416\, d^2$

$r = 0.6204 \times$ cube root of V

Spherical sector

V = volume; A = total area of conical and spherical surface.

$V = 2.0944\, r^2 h$

$A = 3.1416\, r\,(2h + \frac{1}{2}\, c)$

$c = 2\sqrt{h\,(2r - h)}$

Spherical segment

V = volume; A = area of spherical surface.

$V = 3.1416\, h^2 \left(r - \dfrac{h}{3}\right)$

$A = 2\pi rh$

$c = 2\sqrt{h\,(2r - h)}$

Spherical zone

V = volume; A = area of spherical surface.

$V = 0.5236\, h\left(\dfrac{3\, c_1^2}{4} + \dfrac{3\, c_2^2}{4} + h^2\right)$

$A = 2\pi rh = 6.2832\, rh$

Barrel

V = approximate volume.

If sides are bent to the arc of a circle:

$V = 0.262\, h\,(2\, D^2 + d^2)$

If sides are bent to the arc of a parabola:

$V = 0.209\, h\,(2\, D^2 + Dd + \frac{3}{4}\, d^2)$

Decimal Equivalents of Standard Fractions

Fraction	Decimal	Fraction	Decimal
1/64	0.015625	33/64	0.515625
1/32	0.03125	17/32	0.53125
3/64	0.046875	35/64	0.546875
1/16	**0.0625**	**9/16**	**0.5625**
5/64	0.078125	37/64	0.578125
3/32	0.09375	19/32	0.59375
7/64	0.109375	39/64	0.609375
1/8	**0.125**	**5/8**	**0.625**
9/64	0.140625	41/64	0.640625
5/32	0.15625	21/32	0.65625
11/64	0.171875	43/64	0.671875
3/16	**0.1875**	**11/16**	**0.6875**
13/64	0.203125	45/64	0.703125
7/32	0.21875	23/32	0.71875
15/64	0.234375	47/64	0.734375
1/4	**0.25**	**3/4**	**0.75**
17/64	0.265625	49/64	0.765625
9/32	0.28125	25/32	0.78125
19/64	0.296875	51/64	0.796875
5/16	**0.3125**	**13/16**	**0.8125**
21/64	0.328125	53/64	0.828125
11/32	0.34375	27/32	0.84375
23/64	0.359375	55/64	0.859375
3/8	**0.375**	**7/8**	**0.875**
25/64	0.390625	57/64	0.890625
13/32	0.40625	29/32	0.90625
27/64	0.421875	59/64	0.921875
7/16	**0.4375**	**15/16**	**0.9375**
29/64	0.453125	61/64	0.953125
15/32	0.46875	31/32	0.96875
31/64	0.484375	63/64	0.984375
1/2	**0.5**	**1**	**1.**

Inches Converted to Decimals of a Foot

Inches	Decimal of a Foot	Inches	Decimal of a Foot	Inches	Decimal of a Foot
1/8	.0104	1/8	.2604	6 1/4	.5208
1/4	.0208	1/4	.2708	6 1/2	.5417
3/8	.0313	3 3/8	.2813	6 3/4	.5625
1/2	.0417	3 1/2	.2917	7	.5833
5/8	.0521	3 5/8	.3021	7 1/4	.6042
3/4	.0625	3 3/4	.3125	7 1/2	.6250
7/8	.0729	3 7/8	.3229	7 3/4	.6458
1	.0833	4	.3333	8	.6667
1 1/8	.0938	4 1/8	.3438	8 1/4	.6875
1 1/4	.1042	4 1/4	.3542	8 1/2	.7083
1 3/8	.1146	4 3/8	.3646	8 3/4	.7292
1 1/2	.1250	4 1/2	.3750	9	.7500
1 5/8	.1354	4 5/8	.3854	9 1/4	.7708
1 3/4	.1458	4 3/4	.3958	9 1/2	.7917
1 7/8	.1563	4 7/8	.4063	9 3/4	.8125
2	.1667	5	.4167	10	.8333
2 1/8	.1771	5 1/8	.4271	10 1/4	.8542
2 1/4	.1875	5 1/4	.4375	10 1/2	.8750
2 3/8	.1979	5 3/8	.4479	10 3/4	.8958
2 1/2	.2083	5 1/2	.4583	11	.9167
2 5/8	.2188	5 5/8	.4688	11 1/4	.9375
2 3/4	.2292	5 3/4	.4792	11 1/2	.9583
2 7/8	.2396	5 7/8	.4896	11 3/4	.9792
3	.2500	6	.5000	12	1.0000

Example: 4 3/8 in. is 0.36458 of a foot.

U.S. Gallons into Cubic Feet

Gallons	Cubic Feet	Gallons	Cubic Feet
1	0.134	300	40.10
2	0.267	400	53.47
3	0.401	500	66.84
4	0.535	600	80.21
5	0.668	700	93.58
6	0.802	800	106.94
7	0.936	900	120.31
8	1.069	1000	133.68
9	1.203	2000	267.36
10	1.337	3000	401.04
20	2.674	4000	534.72
30	4.010	5000	668.40
40	5.347	6000	802.08
50	6.684	7000	935.76
60	8.021	8000	1069.44
70	9.358	9000	1203.12
80	10.694	10000	1336.81
90	12.031	50000	6684.03
100	13.368	100000	13368.06
200	26.736	500000	66840.28

Example

Convert 4321 U.S. gallons into cubic feet.

Solution

From table above, find and add conversion figures for 4000, 300, 20, and 1 U.S. gallons. Thus, 534.72 + 40.10 + 2.67 + 0.13 = 577.62 cu. ft.

Cubic Feet into U.S. Gallons

Cubic Feet	Gallons	Cubic Feet	Gallons
0.1	0.75	30	224.4
0.2	1.50	40	299.2
0.3	2.24	50	374.0
0.4	2.99	60	448.8
0.5	3.74	70	523.6
0.6	4.49	80	598.4
0.7	5.24	90	673.2
0.8	5.98	100	748.1
0.9	6.73	200	1496.1
1.0	7.48	300	2244.2
2.0	14.96	400	2992.2
3.0	22.44	500	3740.3
4.0	29.92	600	4488.3
5.0	37.40	700	5236.4
6.0	44.88	800	5984.4
7.0	52.36	900	6732.5
8.0	59.84	1000	7480.5
9.0	67.32	5000	37402.6
10.0	74.81	10000	74805.2
20.0	149.61	50000	374025.9

Example

Convert 555.5 cubic feet into U.S. gallons.

Solution

From table above, find and add conversion figures for 500, 50, 5, and 0.5 cubic feet. Thus, 3740.3 + 374.0 + 37.4 + 3.74 = 4155.44 U.S. gallons.

SYMBOLS FOR PIPE FITTINGS COMMONLY USED IN DRAFTING

Symbols courtesy of Mechanical Contractors
Association of America, Inc.

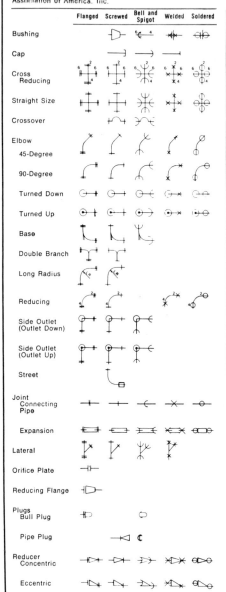

	Flanged	Screwed	Bell and Spigot	Welded	Soldered
Bushing					
Cap					
Cross Reducing					
Straight Size					
Crossover					
Elbow					
45-Degree					
90-Degree					
Turned Down					
Turned Up					
Base					
Double Branch					
Long Radius					
Reducing					
Side Outlet (Outlet Down)					
Side Outlet (Outlet Up)					
Street					
Joint Connecting Pipe					
Expansion					
Lateral					
Orifice Plate					
Reducing Flange					
Plugs Bull Plug					
Pipe Plug					
Reducer Concentric					
Eccentric					

	Flanged	Screwed	Bell and Spigot	Welded	Soldered
Sleeve					
Tee Straight Size					
(Outlet Up)					
(Outlet Down)					
Double Sweep					
Reducing					
Single Sweep					
Side Outlet (Outlet Down)					
Side Outlet (Outlet Up)					
Union					
Angle Valve Check, also Angle Check					
Gate, also Angle Gate (Elevation)					
Gate, also Angle Gate (Plan)					
Globe, also Angle Globe (Elevation)					
Globe (Plan)					
Automatic Valve By-Pass					
Governor-Operated					
Reducing					
Check Valve (Straight Way)					
Cock					
Diaphragm Valve					
Float Valve					
Globe Valve					

	Flanged	Screwed	Bell and Spigot	Welded	Soldered
Gate Valve*					
Motor-Operated					

*Also used for General Stop Valve Symbol when amplified by specification.

	Flanged	Screwed	Bell and Spigot	Welded	Soldered
Motor-Operated					
Hose Valve, also Hose Globe					
Angle, also Hose Angle					
Gate					
Globe					
Lockshield Valve					
Quick Opening Valve					
Safety Valve					

ABBREVIATIONS

ANSI — American National Standard Institute

ASTM — American Society for Testing and Materials

CPVC — Chlorinated Poly (Vinyl Chloride) plastic or resin

IAPMO — International Association of Plumbing and Mechanical Officials

ISO — International Standards Organization

NSF — National Sanitation Foundation

PP — Polypropylene plastic or resin

PPI — Plastics Pipe Institute

PS — Product Standard when reference to a specification for plastic pipe and fittings. These specifications are promulgated by the U.S. Department of Commerce and were formerly known as Commercial Standards.

psi — Pounds per square inch

PVC — Poly (Vinyl Chloride) plastic or resin

PVDF — Poly (Vinylidene Fluoride) plastic or resin

SPI — The Society of the Plastics Industry, Inc.

H

Appendix

BIBLIOGRAPHY

Articles and Bulletins

Atherton, Martin J., "The Assault on PVC Pipes: The Merchandising of Fear," American Council on Science and Health, New York, New York, Sept./Oct., 1981.

"Commonly Used Plastics Industry Abbreviations," Plastics World 1987 Directory

"Fire Characteristics of Rigid Vinyl," B. F. Goodrich Chemical Group, Technical Service Bulletin No. 11, Cleveland, Ohio, 1985

Flax, Steven, "The Dubious War on Plastic Pipe," Fortune, Time, Inc., New York, New York, 1983

Merrick, Ronald C., "A Guide to Selecting Manual Valves," Chemical Engineering/September 1, 1986

"The Permeation of Potable Water Piping Systems," Special Report by PPI Ad Hoc Committee on Permeation (written by Alan J. Olson), New York, New York, 1986

"Pipe Demand to 5 Billion Pounds by 1991," Plastics World/May 1987

"Plastic Pipe in Fire Resistive Construction," Plastic Pipe and Fittings Association, Glen Ellyn, Illinois, 1985

"Plastics under Fire—Accusations' Realities," Plastic Piping Systems, Cranford, New Jersey, 1982

"Specifying Plastic Pipe," Consulting Specifying Engineer/March 1987

Tuthill, A. H., "Installed Cost of Corrosion-Resistant Piping," (parts 1–5), Chemical Engineering/March, April, May, June, 1986 September/October, 1981

Books

ASTM Annual Book of ASTM Standards (1987), ASTM, Philadelphia, Volumes 08.03 and 08.04, 1987

Bliesner, Ron D., *Designing, Operating and Maintaining Pipe Systems Using PVC Fittings (A Handbook of Design Guidelines and Precautions).* Keller-Bliesner Engineering, Logan, Utah, 1986

Chemical Resistance Guide, NIBCO INC., Elkhart, Indiana, 1987

Cheremisinoff, Nicholas P. and Paul N., *Fiberglass-Reinforced Plastics Deskbook,* Ann Arbor Science Publishers, Inc., Ann Arbor, Michigan, 1978

Dym, Joseph B., *Product Design with Plastics,* Industrial Press, New York, New York, 1982

Engineering and Specifications Manual, Eslon Thermoplastics, Charlotte, North Carolina, 1985

Fire Protection Guide on Hazardous Materials, 7th ed., National Fire Protection Association, Quincy, Massachusetts, 1978.

+GF+ Plastic Systems, +GF+ Plastic Systems, Inc., Tustin, California, 1983

Handbook of PVC Pipe, Uni-bell PVC Pipe Association, Dallas, Texas, Second Edition 1982

Harper, Charles A., *Handbook of Plastics and Elastomers,* McGraw-Hill Book Co., New York, New York, 1975

Hilado, Carlos J., *Flammability Handbook for Plastics,* 3rd ed., Technical Publishing Company, Westport, Connecticut, 1982

Janson, Lars-Eric, *Plastic Pipe in Sanitary En-*

gineering, Celanese Piping Systems, Columbus, Ohio, 1974.

King, Reno C., *Sabin Crocker Piping Handbook,* 5th ed., McGraw-Hill Book Company, New York, New York, 1973.

Landrock, Arthur H., *Handbook of Plastics Flammability and Combustion Toxicology,* Noyes Publications, Park Ridge, New Jersey, 1983

Lindsey, Forrest R., *Pipefitters Handbook,* 3rd ed., Industrial Press, New York, New York, 1967

Modern Plastics Encyclopedia, 1986–1987, Vol 63 No. 10A, New York, New York, McGraw-Hill Book Company, 1986.

Mruk, Stanley A., *Engineering Basics of Plastics Piping,* (reprinted with permission) from the *Standard Handbook of Plant Engineering,* by McGraw-Hill, New York, 1983

NIBCO/CHEMTROL Plastics Piping Handbook, NIBCO INC., Elkhart, Indiana, 1984

Plastic Piping Institute Technical Reports (see following page)

Plastic Piping Manual, PPI, New York, New York, 1976

Pro Line®, Super Pro Line® Engineering Design Guide, Asahi/America, Medford, Massachusetts, 1986

Process Piping Design, Scepter Manufacturing Company, Ltd., Ontario, Canada, 1986

Schweitzer, Philip A., *Handbook of Corrosion Resistant Piping,* New York, Industrial Press, Inc., 1969

Thermoplastic Piping Systems Technical Manual, R & G Sloane Mfg. Co., Sun Valley, California, 1983

Catalogs

Asahi/America—Plastic Piping Products, Medford, Massachusetts, 1986

Bristol Pipe, Inc.—PVC Pipe, Bristol, Indiana, 1986

Carlon—PVC Sewer Pipe, Cleveland, Ohio, 1985

Conley Corporation—Fiberglass Piping, Tulsa, Oklahoma, 1986

Dore'—Lined Pipe, Houston, Texas, 1979

Dupont—Polyethylene Pipe, Wilmington, Delaware, 1987

Dupont—Teflons, Wilmington, Delaware, 1987

Duraplus (Div. of Enfield Industrial Corp.)—ABS Piping, Northbrook, Illinois, 1987

Enfield Industrial Corp.—Waste Piping Systems, Northbrook, Illinois, 1987

Eslon Thermoplastics—Plastic Piping Products, Charlotte, North Carolina, 1986

Fibercast (Div. of LTV Energy)—Fiberglass Piping, Little Rock, Arkansas, 1987

Flo Control, Inc.—Plastic Flow Devices, Burbank, California, 1987

+GF+ Plastic Systems, Inc.—Plastic Piping Systems, Tustin, California, 1987

H & W Industries, Inc.—PVC Pipe, Booneville, Mississippi, 1983

Harvel Plastics, Inc.—PVC/CPVC Products, Easton, Pennsylvania, 1983

Hayward Mfg. Company—Plastic Valves/Strainers, Elizabeth, New Jersey, 1987

NIBCO/CHEMTROL—Plastic Piping Products, Elkhart, Indiana, 1987

PSI—Lined Pipe, Wilmington, Delaware, 1985

Pennwalt—Kynar, Philadelphia, Pennsylvania, 1986

Phillips Driscopipe, Inc.—Polyethylene Pipe, Richardson, Texas, 1987

Plast-o-matic Valves, Inc.—Plastic Valves, Totowa, New Jersey, 1987

Poly-Technology (Granse Co.)—Polyethylene Pipe, Lakeville, Minnesota, 1986

Rahn Corporation—Polybutylene Pipe, Ontario Canada, 1987

Sani-Tech—Plastic Sanitary Processing Systems, Andover, New Jersey, 1987

Spears Mfg. Company—Plastic Piping Products, Sylmar, California, 1986

Vanguard Plastics, Inc.—Polybutylene Pipe, McPherson, Kansas, 1987

Plastic Piping Institute—Technical Reports

TR3-86 Policies and Procedures for Developing Recommended Hydrostatic Design Stresses for Thermoplastic Pipe Materials

TR4-86 Recommended Hydrostatic Strengths and Design Stresses for Thermoplastic Pipe and Fittings Compounds

TR5-86 Standards for Plastic Piping

TR7-68 Recommended Method for Calculation of Nominal Weight of Plastic Pipe

TR8-68 Information Procedures for Polyethylene (PE) Plastic Pipe (1968)

TR9-81 Recommended Services (Design) Factors for Pressure Applications of Thermoplastic Pipe Materials

TR10-69 Recommended Practice for Making

Solvent Cemented Joints with Polyvinyl Chloride Plastic (PVC) Pipe and Fittings

TR11-69 Resistance of Thermoplastic Piping Materials to Micro- and Macro-Biological Attack

TR13-73 Polyvinyl Chloride (PVC) Plastic Piping Design and Installation

TR14-71 Water Flow Characteristics of Thermoplastic Pipe

TR15-73 Recommended Practice for Bending Polyvinyl Chloride (PVC) Conduit in the Field

TR16-73 Thermoplastic Water Piping Systems

TR17-72 Thermoplastic Piping for Swimming Pool Water Circulation Systems

TR18-73 Weatherability of Thermoplastic Piping

TR19-84 Thermoplastic Piping for the Transport of Chemicals

TR20-73 Joining Polyolefin Pipe

TR21-86 Thermal Expansion and Contraction of Plastic Pipe

TR22-74 Polyethylene Plastic Piping Distribution Systems for Components of Liquid Petroleum Gases

TR23-74 General In-Plant Quality Control Program

TR24-75 Deflection of Thermoplastic Pipe Resulting from Thermal Cycling

TR27-75 Thermoplastic Drainage Systems for Residential Applications

TR28-76 Installation Procedures for PVC High Pressure Irrigation Piping Systems

TR30-76 Thermoplastics Fuel Gas Piping, Investigation of Maximum Temperatures Attained by Plastic Pipe Inside Service Risers

TR31-79 Underground Installation of Polyolefin Piping

Plastic Piping Institute—Technical Notes

TN2-70 Sealants for Polyvinyl Chloride (PVC) Plastic Piping

TN3-71 Electrical Grounding

TN4-71 Odorants in Gas Pipelines

TN5-72 Testing Equipment

TN6-72 Recommendations for Coiling Polyethylene Plastic Pipe and Tubing

TN7-73 The Nature of Hydrostatic Time-To-Rupture Plots

TN8-73 Making Threaded Joints with Thermoplastic Pipe and Fittings

TN9-73 Coiling PVC Pipe and Tubing

TN10-75 Descriptions of Plastic Piping Joints

TN11-77 Suggested Temperature Limits for Thermoplastic Piping in Non-Pressure Applications

TN12-77 Coefficients of Thermal Expansion Thermoplastic Piping Materials

TN13-81 General Guidelines for the Heat Fusion on Unlike Polyethylene Pipes and Fittings

TN14-86 Plastic Pipe in Solar Heating

Plastic Piping Institute—Recommendations and Statements

Recommendation A Limiting Water Velocities in Thermoplasic Piping Systems (1971)

Recommendation B Thermoplastic Piping for the Transport of Compressed Air or Other Compressed Gases (1972)

Recommendation C Pressure Rating of PVC Plastic Piping for Water at Elevated Temperatures (1973)

Statement D Polyethylene Plastic Pipe Systems for Commercial Propane Gas Distribution (1973)

Statement E Criteria for Joining Various Polyethylene Materials to One Another by Heat Fusion Technique (1974)

Statement F Crush Strength and Flexibility of Thermoplastic Piping (1974)

Statement G Indented Markings (1974)

Statement H Noise in Piping (1975)

Statement I Essential Aspects of the Insurance Services Office (ISO) Grading Systems (1975)

Statement J Hydrostatic Strengths and Suggested Pressure Ratings for CPVC 4120 at Various Temperatures (1976)

Recommendation K Standard Quality Control Form (1977)

Statement L Thermoplastic Piping in Fire Sprinkler Systems (1979)

Statement M Proof Testing of Thermoplastic Pipe (1978)

Statement N Pipe Permeation (6/84)

Statement O PE Materials for Closed Loop Water Source Heat Pump Earth Coils (2/86)

Index

*Plastic pipe on cover provided by National Molded, Inc.,
Jersey City, NJ.
Book typography, cover photograph, and design by Stevan A. Baron.